Agriculture and Public Goods

Francesco Vanni

Agriculture and Public Goods

The Role of Collective Action

Springer

Francesco Vanni
Instituto Nazionale di Economia Agraria (INEA)
Rome
Italy

ISBN 978-94-007-7456-8 ISBN 978-94-007-7457-5 (eBook)
DOI 10.1007/978-94-007-7457-5
Springer Dordrecht Heidelberg London New York

Library of Congress Control Number: 2013948476

Printed on acid-free paper

Springer is part of Springer Science+Business Media (www.springer.com)

From a centralistic, technocratic perspective a landscape of local and regional institutions which were set up to deal with local problems are often regarded as "chaotic". But this labeling is wrong. In fact, the capacity of associations set up by responsible citizens to find solutions for real problems is outstanding and more human.

Elinor Ostrom
(*Nobel Prize Lecture*, 2009)

Introduction

The public support to agriculture is increasingly under scrutiny from governments, academics, policy analysts, NGOs and producers/organizations. In particularly, at the EU level, the debate on the future orientation of the Common Agricultural Policy (CAP) is increasingly shaped by the role of agriculture in providing public goods, and there is a broad consensus that this approach will be particularly relevant in legitimating the EU policy intervention in agriculture in the future.

During the last decade, an increasingly number of studies have analyzed the complex relationships between agriculture and public goods, describing all the positive or negative effects (externalities) associated with agriculture for which markets are absent.

The classifications of public goods associated to agriculture usually identify two main categories: environmental goods and non-environmental goods (or social goods). In the first category are placed those public goods closely related to environmental externalities, such as farmland biodiversity, water availability and quality, resilience to flooding and fire, climate stability (mainly carbon storage and reducing greenhouse gas emission), agricultural landscape. In the second category are placed those public goods more related to the social dimension of agricultural activities, such as farm animal welfare and health, rural vitality and food security.

From a theoretical perspective, the provision of public goods and the provision of positive environmental externalities through agriculture have been acknowledged by the concept of multifunctional agriculture. The well-known OECD's (2001) working definition of multifunctional agriculture includes two core elements that are particularly relevant for the public goods debate, both in terms of theoretical development and policy definition. The OECD acknowledges 'the existence of multiple commodity and non-commodity outputs that are jointly produced by agriculture and the fact that some of the non-commodity outputs exhibit the characteristics of externalities or public goods, with the result that markets for these goods do not exist or function poorly'.

Thus, multifunctionality has represented the main conceptual framework that has been used, in both the institutional and academic debate, to describe the public goods objectives that have been pursued through the EU agricultural policies, especially during the last decade.

In order to incentivize the multifunctional role of agriculture and to overcome the market failures caused by public goods and externalities associated to the primary sector, a broad set of agri-environmental policies are currently in place, based on environmental standards and regulation, taxes, payments, tradable permit schemes, etc.

An increasing number of scholars are analyzing the effectiveness of these measures, while other studies focus on the proposals on how to re-orient agricultural policies to increase their efficiency and effectiveness in providing public goods. All these studies show that no single instrument can achieve all the public goods objectives, but in many cases policy mixes are needed in order to combine instruments that complement each other.

At the same time the literature shows that in many cases the policy tools implemented to date have been largely inadequate to provide agri-environmental public goods at the required scale, by acknowledging the need of carrying out additional theoretical and empirical researches on the relations between agricultural and rural development policies and the provision of agri-environmental public goods.

It may be also observed that, in this academic and institutional debate, one of the more understudied issues is the role of collective action for the provision of agri-environmental public goods.

Indeed, while the many policy tools related to public goods associated with agriculture have mainly focused on individual farms rather than on collective action, in many cases it is evident that in order to provide effectively agri-environmental public goods such as biodiversity and landscape a collective approach is necessary, with a joint involvement of farmers and of other rural stakeholders in the same area.

It may be also observed that the majority of studies related to natural resources and collective action are mainly based on the management of Common Pool Resources (CPRs) in developing countries, while few studies are focused on general collective action theory or on collective action for agri-environmental public goods in developed countries.

At the EU level, for example, it is not clear to what extent collective action could be taken into consideration as a valuable alternative to market or state regulation in contributing to the provision of environmental public goods associated with agriculture, and to what extent it is possible to design and implement agricultural policies that incorporate a collective and collaborative approach between different stakeholders in rural areas.

The aim of this book is to analyze the role of collective action in the provision of agri-environmental public goods, and in reducing environmental externalities through agriculture. In addition, the book aims at proving policy recommendations regarding agri-environmental strategies that, especially in the context of the EU policies, could be based on these collective approaches.

Could collective action be an important driver of workable solutions regarding the provision of agri-environmental public goods through agriculture? Which are the main drivers that stimulate the participation of farmers into collective action for public goods? How can government stimulate collective action? At which level of government collective action can it be better stimulated? What kinds of policies are necessary to promote collective action?

The book aims at addressing these questions by exploring, through participatory methods, two case studies of collective action for the provision of agri-environmental public goods that have been recently developed in Central Italy.

The in-depth analysis of these case studies was completed with an extensive literature review, in order to contribute some insights to the theoretical development on the role of collective action for the public goods associated to agriculture. Indeed, as observed above, this argument, at least in the context of developed countries, is largely unexplored.

The book is divided into seven chapters.

The first chapter describes the theoretical approach regarding the definition and provision of public goods. This theoretical approach, rooted in institutional economics, is then applied to the relation between agriculture and the associated public goods and externalities. The chapter describes the different approaches to multifunctional agriculture as well as the type of analytical framework needed to address the public goods agenda in a policy context. It is shown how the different approaches in the notion of multifunctionality have different implications in terms of policy development and policy implementation. This analysis critically assesses the most conventional scientific formulations for approaching multifunctionality, based on neoclassical economics, by showing how this approach provides only partial and limited insights into possible trajectories of the public intervention in agriculture. It is then shown that the conventional approach has usually a narrow focus on the policy instruments needed to incentivize multifunctional activities at farm level, while a 'wider' approach on multifunctionality, rooted in institutional economics, allows the exploration of innovative policy tools and institutional arrangements that may be effective in providing public goods at a territorial scale.

On the basis of this theoretical approach, chapter two focuses on the definitions, role and characteristics of collective action. A very relevant issue to consider when analyzing the dynamics of collective action is what type of organization has developed such action, and this section provides a detailed description of the main institutional arrangements that may favour the development of grass roots collective action aimed at increasing the provision of public goods. It is argued that some innovative institutional arrangements based on mixed private-public solutions, such as co-production and co-management, may represent effective territorial strategies to promote and support collective action. The analysis then is shifted towards the new structure of the agricultural knowledge and innovation systems in agriculture necessary to favour the implementation of collective strategies aimed at providing public goods.

Chapter three provides a review of examples of collective agri-environmental initiatives that have been developed in different contexts, by showing their relevance in the current policy and academic debate. The description of these strategies provides some insights on the potentialities and barriers to implementing collaborative agri-environmental actions in different contexts. The final part of the chapter focuses on the main policy and institutional challenges for developing and supporting grass roots collective actions for public goods by providing a short overview of the Italian case studies that are analyzed more in depth in chapters four and chapter five.

Chapter four describes the collective action related to the 'Custody of the Territory' project, an initiative promoted by a territorial agency of a mountain area of Tuscany ('Media Valle del Serchio') which set an agreement with local farmers for co-production of some environmental services such as the cleaning of rivers, riverbeds, rivers banks and canals. This collective action shows how the relations between farming activities and environmental services could be addressed at a territorial level, which resources are mobilized, what type of information is exchanged and what outcomes are reached. Finally this case study shows how social learning and co-production of knowledge (amongst farmers, institutions, technicians, and citizens) are very important issues for a collective provision of environmental services.

Chapter five focuses on the innovative and collective approach to agri-environmental action which was experienced in the Valdaso area (Marche region), where a group of farmers started a grass roots initiative to adopt integrated agriculture at a territorial scale with the objective of protecting water and soils from pesticide and nitrate pollution. This collective action was supported by the regional and provincial authorities, which settled a territorial agri-environmental agreement financed by the regional Rural Development Programme. The case study shows that farmers' collective action may play a significant role in controlling negative externalities from agriculture, especially if local institutions positively influence collective decision making behaviors, by structuring a range of incentives, capacity building programs and technical assistance to align individual and group interests and, above all, private and public goods objectives. The description of this initiative allowed me to explore, in the last section of the chapter, how a collective approach to agri-environmental action could be better embedded and institutionalized in the current political settings.

Chapter six summarizes the findings of chapters four and five and provides a more detailed theoretical discussion on the role of collective action in the provision of agri-environmental public goods. In this chapter the policy implications of the main results are discussed, also in the light of the current debate on the policy and institutional innovation needed to re-orient the EU Common Agricultural Policy towards a more effective provision of public goods in rural areas.

Chapter seven draws some conclusions regarding the role of collective action in agriculture, together with some suggestions for further research.

Acknowledgement

First, I would like to thank Gianluca Brunori for his encouragement and constant help from the initial development of my ideas, through to the book's completion. Special thanks also go to Gilles Allaire, for his invaluable guidance regarding the theoretical part of the research.

Thanks also to Massimo Rovai and Simone Gorelli, who have both provided me with information, contacts and material for the case studies.

Many thanks to Silvia Coderoni, for her valuable advice but also for her encouragement and useful critiques of my research work.

I would also like to extend my thanks to the technicians, farmers and representatives of local and regional institutions who provided me with information, contacts and material on the case studies; especially Gianfranco Vagnoni, Ilaria Ferri, Pamela Giani and Ferruccio Carrara.

I also thank Mark Poole, for his editing work on my written English.

A very special mention must be given to Valeria Rinaudo, who had confidence in me and who gave me the opportunity to write this book.

Finally, it would have been impossible to complete this book without Veronica and Matteo's love, support, and encouragement. This book is dedicated to them.

Contents

List of Abbreviations

AEA	Agri-Environmental Agreement
AES	Agri-environmental Schemes
AIS	Agriculture Innovation System
AKS	Agricultural Knowledge System
ASSAM	Agenzia Servizi Settore Agroalimentare delle Marche
CAP	Common Agricultural Policy
CPRs	Common Pool Resources
DVL	Deutscher Verband für Landschaftspflege (German Association for Landcare)
GAEC	Good Agricultural and Environmental Condition
GT	Grounded Theory
ILDS	Integrated Local Development Strategies
IPM	Integrated Pest Management
ITP	Integrated Territorial Project
LAG	Local Action Group
LCA	Landcare Association
LRDP	Local Rural Development Plans
MD	Mating Disruption
NLP	National Landcare Programme
NVZ	Nitrate Vulnerable Zone
PRA	Participatory Rural Appraisal
PES	Payments for Environmental Services
RDP	Rural Development Plan
RRA	Rapid Rural Appraisal
UAA	Utilised Agricultural Area

Chapter 1
Agriculture and Public Goods

Abstract The chapter describes the theoretical approach undertaken in the book to define and analyze public goods. This theoretical approach, rooted in institutional economics, is then applied to the relationship between agriculture and the associated public goods and externalities. Since the public goods provision through agriculture is strongly related to the concept of multifunctionality, the chapter also provides an overview of the different approaches to multifunctional agriculture as well as the type of analytical framework needed to address the public goods agenda in the EU policy context. This analysis shows how the different approaches in the notion of multifunctionality have different implications in terms of policy development and policy implementation. The chapter critically assesses the most conventional scientific formulations for approaching multifunctionality, based on neoclassical economics, by showing how this approach provides only partial and limited insights into possible trajectories of the public intervention in agriculture. Whilst the conventional approach has usually involved a narrow focus on the policy instruments needed to incentivize multifunctional activities at farm level, a 'wider' approach on multifunctionality, rooted in institutional economics, enables the exploration of innovative policy tools and institutional arrangements that may be effective in promoting a collective and territorial approach to public goods provision.

Keywords Institutional economics · Public goods · Multifunctional agriculture · Common agricultural policy

1.1 Institutional Approach to Public Goods Provision

According to the well-known definition of Samuelson (1954), the main characteristics of public goods are non-excludability and non-rivalry. A good is public when if it is available to one person, others cannot be excluded from the benefits it confers (non-excludable); at the same time, if a public good is consumed by one person it does not reduce the amount available to others (non-rival). These characteristics imply that users have no incentive to pay for the consumption of such goods and, on the supply side, there are no incentives for providing public goods, because potential producers are not remunerated by the market to do so. The combination of these factors explains the so-called 'market failure', and the reason for the need of public intervention in order to achieve a socially optimal level of public goods, consistent with societal demand.

F. Vanni, *Agriculture and Public Goods*,
DOI 10.1007/978-94-007-7457-5_1, © Springer Science+Business Media Dordrecht 2014

Table 1.1 Typologies of goods (Source: Author elaborations on Cooper et al. (2009) and OECD (2001))

Level of 'publicness'	Type of good	Excludability and rivalry	Example
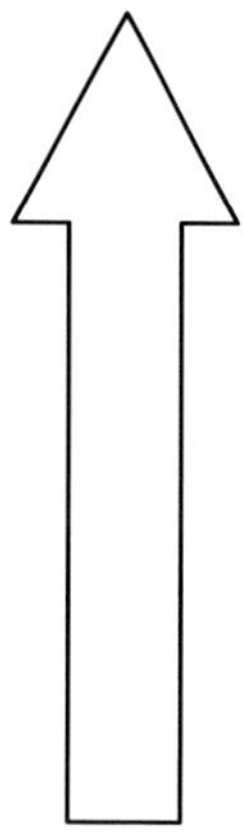	Pure public goods	Non-rival Non-excludable	Biodiversity Non-use values of land-scape
	Impure public goods (Common Pool Re-sources)	Non-rival Excludable only at high costs (high risk of conges-tion)	Public access to farm-land Ground and surface wa-ter Soil conservation
	Club goods	Non-rival for a small user group Excludable (subject to congestion)	Private parks Golf course
	Private goods	Rival Excludable	Wheat, timber

According to this approach, the reasons beyond the so-called market failure are related to the characteristics of non-excludability and non-rivalry, which determine both the lack of incentives to produce public goods and the presence of opportunistic behaviors (free riding). The efficiency of market mechanisms regarding the allocation of goods is mainly related to the characteristic of (non) excludability, since market mechanisms work better for goods with high-excludability levels (Merlo et al. 1999). At the same time, goods with a low-excludability level present several problems related to congestion and over-exploitation, as described in the 'Tragedy of Commons' (Hardin 1968).

In case of market failure, public intervention is needed to avoid the under provision of public goods, which is determined by the degree of non-excludability and non-rivalry of goods and, as consequences, on their degree of 'publicness'. Indeed, by using the concepts of non-rivalry and exclusion, it is possible to show that there are intermediate forms between pure public goods and private goods. Common goods or common pool resources are the goods where rivalry exists but exclusion is not possible (e.g. common fish grounds or water systems) and quasi-public goods are the goods (also called club goods or toll goods) where exclusion is possible, but rivalry does not exist (see Table 1.1).

The conventional approach based on the Samuelson definition presents several conceptual and operational limitations, as highlighted by several economists who have carried out their analysis on public goods provision through an institutional approach (Kaul and Mendoza 2003; Hagedorn 2008).

According to these authors, from a theoretical perspective the limitations of this conventional approach are mainly related to the narrow definition of private/public goods. While the definition of public goods based on (non)excludability and the (non)rivalry is based solely on market criteria, the private and public domains of

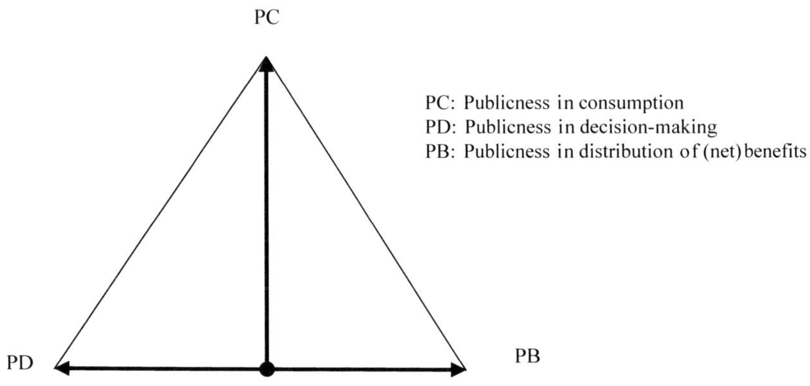

Fig. 1.1 The triangle of 'publicness'. (Source: Kaul and Mendoza (2003))

goods in reality are also determined by the general public and by the political process. Thus, it is necessary to consider goods not only in their original forms, but also as social constructs and as results of deliberative policy choices (Kaul and Mendoza 2003). According to this vision, public goods are not just market failures, since public and private domains exist on their own, but are those goods, technically non excludable, which are placed or left in the public domain by policy choice.

While the definition of private/public goods based on non-excludability and the non-rivalry is an indirect definition based on market conditions (private/public are mainly considered as market/non-market goods), in reality such characteristics do not specify properties of goods themselves, but refer to the institutional and organizational fit of a social construction (Hagedorn 2008).

Kaul and Mendoza (2003) observed that, together with the evolution of social institutions, many goods have developed into mixed cases, showing both exclusive and non-exclusive proprieties that have a temporary dimension, since they may change, for example, as result of the development of new technologies and, above all, according to the different policies and regulation that are implemented.

By introducing the notions of social construction and public choice in the definition of public goods, these authors provide a useful conceptual framework for evaluating the publicness of goods according to this perspective. Through the triangle of publicness (Fig. 1.1), it is possible to examine goods according to three dimensions: (i) publicness in decision-making, (ii) publicness in the distribution of a good's benefits and (iii) publicness in consumption.

The first criteria to assess the publicness of goods are represented by the base of the triangle:

- The publicness of decision-making (PD) is the criterion used to assess the participatory nature of the process to place goods in the public domain and also refers to the decisions related to the provision of the good, such as the level and the modalities of production of the goods and how to distribute its benefits among the users;

- The publicness in distribution of benefits (PB) is the criterion used to assess the equity of benefits, namely to what extent the various groups of users and consumers derive the benefits from the public goods;

These two criteria have to be combined with the main criterion to assess the publicness of goods:

- The publicness of consumption (PC), which represents the criterion to assess the non-exclusiveness across individual and groups.

The triangle of publicness provides a theoretical foundation for examining the discrepancies between the publicness of a single good in the three dimensions. Indeed, while Fig. 1.1 shows an ideal situation where a good fully meets the three conditions, usually goods do not fully meet all the three criteria. Nevertheless, this conceptualization may help policymakers and the public to understand the issues to be addressed to increase the provision of public goods though policy tools, as well as through an institutional change and new governance settings.

From a theoretical point of view it may be argued that this extended definition of public goods allows going beyond the incomplete market/state dichotomy. Traditionally neo-classical economics have dealt with the provision of market goods by private actors in opposition to the need for public management for goods with some degree of 'publicness'. This approach has been unsatisfactory from many points of views and especially for goods with intermediate features between public and private goods. Indeed, the two criteria of (non)excludability and the (non)rivalry are not helpful for deciding what governance structures are suitable for coordinating transactions of goods that do not fulfill the criteria: they only provide a general indication that these goods should be dealt with by any other governance structure (Hagedorn 2008).

Many environmental resources, such as forests, pastures and fisheries are common-pool resources that in some circumstances are still governed by customary common property arrangements. On the other hand, there are also some larger common-pool resources (bodies of water, air basins, and global atmosphere) that are usually governed by formal governance institutions. These types of goods and resources do not involve only different types and entities of public intervention, but, above all, create different kinds of governance challenges. Starting from this consideration, new theoretical and analytical approaches have been developed in order to explore innovative institutional arrangements and governance settings for the provision of public goods and common goods resources. In particular, Ostrom, in her extensive work, developed a new theoretical and analytical framework for the management of collective goods and services, by exploring the management of commons goods with the direct involvement of local communities regarding both the allocation and the use/exploitation of the common resources.

Through a neo-institutional approach, Ostrom has focused her analysis on the local dimension, in order to observe the process of self-organization of local communities and on the ways in which the subjects who are in a condition of interdependency can obtain permanent results, even though are tempted of over-exploiting the resources.

According to Ostrom (1990), in order to make rules effective it is necessary for there to be a presence of organizational solutions (governance structures), in order to guarantee the rights and the duties and their use in coordinating transactions. Governance structures, which may be public or private forms of organizations, includes hierarchies, markets, hybrid forms, planning processes, knowledge and information systems and networks, monitoring infrastructures, procedures for conflict resolution and distribution of costs, and incentives to promote innovation and learning (Prager 2010, p. 228).

This approach, by involving a broader conceptualization of public goods, provides a different analytical framework for the definition of policy tools for public goods provision. This model clearly shows the need to go beyond the criterion of cost efficiency alone and the need to add equity considerations both for the supply and for the demand side of public goods provision.

Thus, this broader conceptualization allows the identification of another important limitation of the conventional approach, which is related to the concept of optimal supply of public goods. As defined by the Bowen-Lindahl-Samuelson condition, the optimal supply requires that the sum of all individuals' marginal willingness to pay for an additional unit of a particular public good equal the marginal cost of producing that unit. Many conceptual and methodological problems arise when attempting to empirically assess this condition. Indeed, especially for global public goods, the mechanisms to be implemented are particularly complex, due to the high number of potential consumers and their heterogeneity in terms of features and proximity to goods (Desjeux et al. 2011). In addition, while this condition provides guidance on how to adapt government revenue and the spending of consumers, a balanced pattern of spending and preferences does not necessarily indicate that a public good is adequately provided (Kaul and Mendoza 2003). Thus, the optimal supply should be substituted by the concept of adequate provision, which is based solely on technical considerations, without reference to costs, benefits or existing preferences and willingness to pay (Conceição 2003). The criterion of adequacy is not meant to indicate optimality but is meant, rather, to establish a simple and reliable tool for measuring the present provision of a certain good against a technical notion of adequacy, which may be defined as the missing or biased components and the actions needed to enhance the current provision of such a good (ODS 2002).

The actions to achieve the adequate provision depend on the type of public good, as well as the related 'corrective actions', which may include new standards and regulations, financing capacity-building measures, subsidies and direct payments. Even though the notion of adequate provision has been mainly debated in relation to the provision of global public goods at international level (the concept was developed in the context of United Nation debate, see Kaul et al. 2003; Conceição 2003 and ODS 2002), this approach may also provide useful insights regarding the debate around the provision of public goods through agriculture at the EU level, where the focus on the cost efficiency of the measures in many cases has resulted to an undersupply of the most relevant public goods associated to the primary sector.

1.2 A 'Wider' Approach on Multifunctional Agriculture

In the context of the European academic and institutional debate, the relations between agriculture and the associated public goods have been acknowledged by the concept of multifunctionality. It is possible to distinguish two analytical approaches on the multifunctionality of agriculture (see Aumand et al. 2006; Van Huylenbroeck et al. 2007): the first one focuses on the supply side issue (positive approach) while the second one on the demand side issue (normative approach).

The supply-side approach is well synthesized by the working definition of the OECD (2001), which defines multifunctionality as 'the existence of multiple commodity and non-commodity outputs that are jointly produced by agriculture and the fact that some of the non-commodity outputs exhibit the characteristics of externalities or public goods, with the result that markets for these goods do not exist or function poorly'. This definition includes the core elements for understanding the relation between the agricultural activities and public goods provision from a supply side perspective, both in terms of theoretical development and in terms of policy definition. Indeed, this approach conceptualizes public good provision in terms of joint outputs of an activity or of a combination of activities, where 'the term multifunctionality refers to this nexus between commodity and non-commodity output production in agriculture' (OECD 2008). The analysis of this *jointness* is at the core of this approach, since this concept involves important implications in terms of public support to the primary sector and on the evaluation of the agricultural policies implemented for the public goods objective. As observed by Vereijken (2001), the level of *jointness* of different public goods with agriculture activities is very complex and is usually characterized by high variability across areas, countries and specific environmental and social goods.

Conversely, the demand-side approach looks at multifunctional agriculture as a societal objective and at the demand of society for the multiple functions of agriculture. This approach involves a concept of multifunctional agricultural more territorially embedded and more related to rural areas rather than to the agricultural sector. Indeed, according to this vision, multifunctional agriculture must satisfy societal expectations and meet societal demand and needs not only in relation to the agricultural sector, but also according to agricultural production processes and the spatial extent of agriculture.

In this approach, the characteristics of non-rivalry and non-excludability of goods are straightened by the 'non-user values' or 'option values', which reflect the interests of citizens in securing the provision of public goods, such as habitat preservation or the protection of endangered species, and who is willing to bear the costs, even though they are not immediate users and the goods concerned may be a long distance from them (ENRD 2010). This approach is usually characterized by economic analyses which aim at assessing the demand of external benefits produced by the agriculture in monetary terms, both through indirect methods (contingent valuation method) and through direct methods (hedonic pricing method) (i.e. see Randall 2002).

According to Van Huylenbroeck et al. (2007), the main difference between the supply vision and the demand vision lies in the implicit treatment of externalities: while the supply definition considers negative and positive externalities as good and bad outputs respectively and treats them equally; the demand approach privileges the positive contributions of agriculture to public goods, usually emphasizing the positive contribution and benefits that agriculture may deliver to society.

However, besides the supply vision and the demand vision, there is also a third and more holistic interpretation of the concept of multifunctionality. According to this 'wider' perspective, multifunctional agriculture is the result of a transformation process in the relationship between agriculture, rural society and society at large. According to this approach, the growing attention to multifunctional agriculture is not considered a direct response to market failure, but rather a consequence of the changing needs and demands of consumers and society in combination with the failure of conventional, productivist farm development models (see Van Huylenbroeck and Durand 2003; Renting et al. 2009; Wilson 2007).

According to this wider vision, new institutional arrangements are becoming more and more relevant in order to achieve important results concerning the new priorities attached to rural areas, since it is necessary to articulate the demand for countryside goods, establishing incentives for resources managers and coordinating resource management across space (Hodge 2001). This emphasis towards the territorial provision of countryside goods does not remove the requirement to promote supply at minimum cost, but it also emphasizes the need for new institutions with the difficult task of providing a diverse range of goods according to the public goods' characteristics and to the range of environments within which they are to be supplied.

Indeed, the different types of public goods that agriculture may produce differ in aggregation, scale and nature, and this claims for different coordination mechanisms amongst institutions (Hagedorn et al. 2002). In some cases institutions and governance structures may be designed intentionally by economic and political actors, while in many cases such structures may evolve unintentionally or be self-organized.

The need for a more integrated approach, which could adopt a more territorial vision of multifunctional agriculture has also been highlighted by the OECD in its publication on the 'new rural paradigm' (OECD 2006) and this model is increasingly influencing the EU academic and institutional debate.

The recent studies on the CAP and public goods provision (Cooper et al. 2009; RISE foundation 2009; ENRD 2010; Hart et al. 2011) have made an effort to integrate some issues related to the demand side approach in the policy analysis, in order to obtain a sound theoretical framework which takes into account in a more exhaustive way both the positive and normative approach on multifunctional agriculture.

With regard to public goods provision, the analytical focus of this approach is not restricted to pure public goods (biodiversity, landscape, sustainable water management), but also includes private goods and services (energy, tourism, social farming) and 'functions' that are provided by agriculture as distinctive product attributes on specific food markets (food quality, animal welfare, organic products etc.). Finally, functions and services that are not directly linked to agricultural production are also considered, such as the vitality of rural areas, rural viability, and the maintenance of settlement patterns in remote rural areas.

According to this wider perspective, the degree of farms multifunctionality is highly influenced by a wide range of—often interrelated—institutions and policies, and for this reason it is necessary to go beyond the livelihood strategies, by analyzing the institutional arrangements necessary to implement successful local actions which enhance the multifunctional role of agriculture. According to this approach, 'jointness of production' is not defined in relation to the production functions, but regarding the economic and institutional arrangements which may deliver a combination of commodities and non-commodities, private and public goods and in many cases may also deliver other types of non-private goods, such as club goods and common-pool resources. Thus, it is necessary to explore how the various local actors (farmers, politicians, bureaucrats, agricultural organizations, environmental NGOs, national and EU institutions) can arrange relevant economic activities related to the primary sector in order to make use of the multifunctional capacities of agriculture (Hagedorn 2004).

The theoretical framework developed by these scholars, based on the study of multifunctional agriculture, by looking at this wider institutional relations of the farm with social networks, markets, consumer groups and policy frameworks (Van der Ploeg 2000; Van der Ploeg and Roep 2003), provides a sound conceptual basis for exploring the role of collective action in the provision of environmental goods associated with agriculture.

1.3 The CAP and the Provision of Public Goods

As described in the previous section, the provision of public goods from agriculture is strongly related to the concept of multifunctional agriculture. This concept emerged at European level during the 1990s as a response of the primary sector to the broader challenge of sustainable development, and it is based on the rediscovering of the additional functions of agriculture besides the production of food and fibre.

The CAP, especially before the MacSharry reform of 1992, was largely criticized for its limited cost-effectiveness and for the largely unforeseen environmental and social consequences of its supporting instruments. Thus, in the context of the most recent CAP reforms—especially in the context of Agenda 2000 in 1999 and of the 2003 Mid Term Review—there has been an increasing effort to implement measures related to the multifunctional role of agriculture, by emphasizing the role of farmers as guardians of the landscape and of natural resources, together with an attempt to better integrate the primary sector in the rural economy.

This strategy, 'officially' implemented to obtain a more integrated and sustainable development of rural areas, has been largely criticized, especially in the extra-EU context. Indeed, several scholars and institutional stakeholders have interpreted the increasing acknowledgment of the secondary functions of agriculture as a tool for legitimizing the implementation of protectionist measures in the EU and, more broadly, for legitimizing the high public support to the agricultural sector, with negative consequences in terms of fair competition within the international markets.

However, the evolution of the CAP has brought forward the development and implementation of strategies that go well beyond an undifferentiated and protectionist support to the sector. On one side, the rural development policy, introduced in the framework of Agenda 2000, has transformed a part of the CAP from a sector policy to a territorial policy which, while not always in a very efficient way, has certainly contributed to the social and economic development of many rural areas across Europe. On the other side, the increasing attention to the agri-environmental measures, together with the implementation of the decoupling and of the cross compliance in the framework of the 2003 reform, were aimed at increasing the integration of environmental issues within the CAP.

Thus, the concept of multifunctionality has represented the main conceptual framework for understanding the social and environmental objectives that have been pursued through these CAP reforms, since the decoupling of direct payments from production, together with the rural development policy, have been conceived as strategic tools to link the EU public support to the objectives of sustainability in a more effective way.

Nevertheless, during the last few years, especially in the institutional and political debate, it seems that the concept of multifunctionality has lost part of its value and the words 'multifunctional agriculture' have been used to a lesser extent. This was particularly evident in the context of the 2008 CAP Heath Check, when the debate on the future of the agricultural policy has been focused, rather than on multifunctional agriculture, on the need to support the public goods and services produced by the agriculture which are not remunerated by the market.

In the debate on the future of the CAP, this transition regarding the political discourse from multifunctional agriculture to public goods provision is particularly significant, since the new words used to define the secondary functions of the primary sector should have involved a new 'normative' vision of agriculture as a producer of immaterial goods. This shift from multifunctionality to public goods provision could have implied, especially amongst the policy makers and the other stakeholders, a broader cultural change which goes well beyond this different 'wording'.

On the opposite side, the word multifunctionality has been abandoned mainly for political reasons, since in the international debate the word had lost its political appeal and it had increasingly been perceived as a way of legitimizing the strong public support for EU agriculture. At the same time, it may be argued that the debate surrounding the new public goods agenda is not characterized by a real innovation regarding the conceptualization of the environmental and social dimensions of the agricultural activities and, above all, by an innovation of the definition of the policy instruments to be implemented to pursue the public goods objective.

Indeed, in spite of the increasing emphasis on public goods provision to define the social and environmental role of agriculture, the current debate is largely based on the main conceptual and policy framework that was used in the debate of multifunctionality of agriculture. As shown above, this conceptual framework is based on neo-classical economics, where the public goods provision is strongly related to the concept of *jointness* of production and of externalities.

The degree to which the recognition and support, through the CAP, of the non-commodity outputs from agriculture have led to an improvement in the environmental performance of the sector is matter of continuous debate. The most controversial issue is disentangling the impacts of specific policy measures from the range of the external drivers affecting agricultural management, but also the variability in the way certain CAP policy measures are funded, designed and implemented nationally and regionally (OECD 2011).

In the current debate, the important role that agriculture has to play in delivering public goods in Europe has been recently recognized in the European Commission's Communication 'The CAP towards 2020: meeting the food, natural resources and territorial challenges of the future' (European Commission 2010). This document sets out three key objectives for the CAP to 2020: viable food production, sustainable management of natural resources and climate action, and balanced territorial development; all three of which relate to the provision of environmental and other public goods by land managers (Hart et al. 2011). The re-orientation of the CAP towards the provision of public goods is supported by several agricultural economists (Anania et al. 2009; Zahrnt 2009) and by the leading EU environmental organizations (Birdlife 2009), but a growing number of producers' organizations also seem receptive to the idea that the CAP budget should be more oriented towards remunerating farmers for the provision of environmental and ecosystem services (Copa-Cogeca 2010).

In the recent proposal on the future of the CAP, the European Commission (2011a; 2011b) proposes several strategies aimed at re-orienting the CAP towards the provision of public goods. In particular, the current debate on the CAP for the 2014–2020 programming period and public goods provision is mainly focused towards the *greening* of direct payments. In the European Commission (2011a) proposal, green payments are articulated as general and 'horizontal measures', targeted on single farms, with the objective of strengthening the approach of cross-compliance.

As it is further discussed below, this approach is still largely based on the remuneration of single farmers for the additional costs and the decreased income derived from the changes of the land management practices, with a strong focus on the policy tools rather than on the social and institutional innovations necessary to implement territorial and collective strategies.

More in detail, at EU level it is possible to identify several policy tools that, to some extent, are related to the provision of environmental goods through agriculture.

Cooper et al. (2009, p. 88) have identified three types of EU policy measures that have some potential to support the provision of environmental public goods:

1. Measures with a direct focus on the provision of public goods: cross-compliance; agri-environmental measures; payments for non-productive investment for farms; LIFE + projects focused on agricultural and structural funds projects under the heading 'Preservation of the environment in connection with land and ... landscape conservation;
2. Measures with a partial focus on the provision of public goods, such as Rural Development advice and training measures; measures supporting farm modernization and infrastructure development; Less Favoured Areas payments; Natura

2000 payments; support for conservation and upgrading of the rural heritage; training and information measures;
3. Measures with no direct focus on the provision of public goods, but that may have a positive impact, such as: direct payments and Rural Development measures on adding value to agricultural products; support for the diversification of farms activities; and encouragement of tourism activities.

Amongst the measures which have been implemented specifically for the provision of environmental public goods (the first group of policy tools above), the most significant in terms of institutional, political and academic interests are the cross compliance and the agri-environmental measures.

The idea beyond the CAP cross-compliance is placing environmental conditions on the receipt of agricultural support payments, an approach that started in the 1990s with the objective of improving the adherence to environmental standards at farm level. While the 1992 and 2000 CAP reforms provided member states with the ability to make the receipt of certain payments conditional on meeting specified environmental standards, cross compliance requirements were made compulsory for member states to apply as part of the 2003 reform of the CAP (OECD 2011).

Implemented since 2005, this mechanism ties EU support for farmers to compliance with specific standards, by penalizing farmers who infringe the law on environmental, public and animal health, animal welfare or land management—by reducing the CAP support they receive.

Cross-compliance mainly covers directives and regulations (SMR—Statutory Management Requirements)—that have existed for years and applies to all farmers but also a set of rules on Good Agricultural and Environmental Condition (GAEC), designed specifically for farmers receiving CAP payments. GAEC standards were designed to promote more sustainable agriculture, as well as to act as a flanking measure to address unwanted side-effects of the introduction of single payments, most notably the cessation of the active management of farmland and the risk of land abandonment (OECD 2011).

In all EU-27 member states, farmers are required to respect certain environmental regulations, whether or not cross-compliance approaches are used. In the EU, due to the introduction of GAEC, the reference level of environmental quality for cross compliance is higher than that defined by the environmental regulations. At the same time, where farmers voluntarily enter into a contract with the government to provide environmental quality beyond what is required (the reference level) and for which no market return exists (public goods), they would be entitled to a compensation or incentive a payment as long as they complied with specified criteria (OECD 2011).

Figure 1.2 shows the relationships between environmental targets, reference levels and farmers' economic optimum (the level of environmental quality farmers would provide on the basis of private profitability) (Fig. 1.2).

With regard to the environmental performance of cross compliance, it has been shown that the introduction of this requirement since 2005 has undoubtedly expanded the area of land subjected to basic environmental management requirements; it has also increased the pressure on a proportion of farmers to comply with baseline

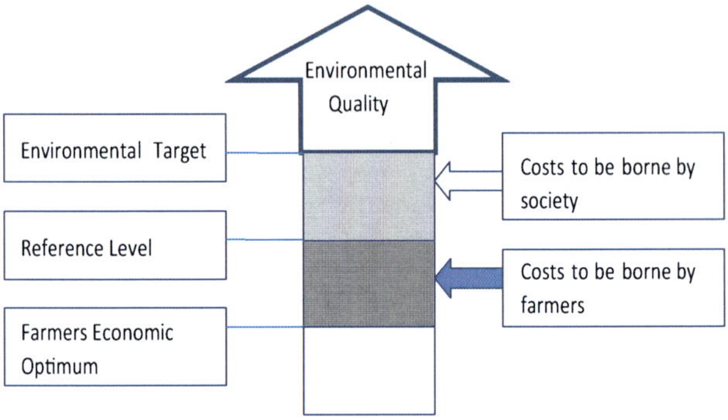

Fig. 1.2 Environmental targets, reference levels and farmers' economic optimum. (Source: OECD (2010))

environmental standards and in many cases raised their awareness of environmental legislation (Hart et al. 2011).

At the same time, several limitations have also been observed, especially in relation to the levels of compliance and enforcement and to the lack of monitoring and evaluation (European Court of Auditors 2008), but also to the general approach of this policy tool.

Indeed, as observed by Brady (2011), targeted measures would be more effective than cross compliance to improve the environmental performance of agriculture, given the immense heterogeneity of agri-environmental conditions across Europe.

Nevertheless, the EU considers this policy tool the most important mechanism for delivering environmental benefits in agriculture across a large land area, and this is confirmed by the recent proposal of the European Commission (2011a) for the CAP for the 2014–2020 programming period.

The current debate on the future CAP is highly concentrated on the *greening* of the first pillar, which consists in three measures to be applied at farm level related to the eligibility of farmers to the 30 per cent in direct payments: (i) crop diversification; (ii) maintaining permanent grassland and (iii) the introduction of ecological focus areas on all eligible arable land per farm.

This strategy confirms that the European Commission considers the approach of cross-compliance as the most cost-effective strategy in delivering environmental goods at large scale, especially in areas characterized by intensive and productive agricultural systems. At the same time, while the greening measures have an annual, non-contractual basis and apply to all farmers in a generic way, public goods such as biodiversity could be better provided through regional coordinated measures that increase landscape heterogeneity through the construction of multi-annual green infrastructures (Westhoek et al. 2012).

The impact of this approach depends very much on local farming systems and on local biophysical and cultural features as well as on the details of the national/local regulations and on the practical implementation at farm level.

From this perspective, higher positive benefits can be reached through local-specific management conditions, especially through the improvement and/or supplement of command and control (and top-down) measures with more targeted rural development measures, in particular agri-environmental schemes.

Agri-environmental schemes are sitting within Pillar 2 of the CAP alongside other land management measures in Axis 2, and they are the oldest and the single most significant measures for pursuing environmental objectives across the farmed landscape, both in terms of the spatial coverage of schemes and the resources allocated to them (OECD 2011).

These measures provide payments to single farmers to adopt specific farming practices on producing land, and are implemented specifically to achieve positive environmental effects and/or providing public goods (such as landscape, biodiversity, etc.). These measures are based on formal contracts between individual farmers and government agencies, under which the farmers agree to follow a particular set of practices and not to undertake others.

Farmers sign a contract with the administration and are paid for the additional cost of implementing such commitments and for any losses of income (e.g. due to reduced production) which the commitments entail. Agri-environmental payments are co-financed by the EU and the Member States and may be designed at national, regional or local level so that they can be adapted to the particular farming systems and environmental conditions (European Commission 2005).

However, these payments in many cases are largely inefficient and they seem inadequate with regard to improving the provision of such goods at the required scale. Their lack of efficiency is mainly related to their lack of targeting and tailoring, but also to the fact that usually their targets are defined in the form of a specific farming practice rather than a specific (measurable) environmental outcome (Vojtech 2010).

Moreover, agri-environmental schemes have generally been promoted through the provision of fixed payments for certain environment-friendly farming or management practices but the location and the quality of agricultural land and production systems used by farmers vary enormously, especially for small households in hilly and mountainous regions, where fixed payments that are uniformly distributed at regional region cannot correspond to the individual heterogeneity (OECD 2012).

Hodge (2001) has identified the main limits of those voluntary policy mechanisms based formal environmental contracts, which may be synthesized in the following points:

- Formal contracts restricts the range of objectives and requirements for farmers, since they involve a strong formalization (common rules and range of actions to be controlled);
- In some cases farmers may evade the contract requirements, because there are always aspects of the contracts which are not readily observable by the government agency;
- The nature of the contracts usually results in a lack of incentives for entrepreneurship and, above all, may limit the incentives and the opportunities for co-operation between landholders.

In addition, while such measures are usually targeted to individual farms, environmental public goods could be more effectively delivered if farmers in a given area take joint action, since in many cases the provision of environmental goods may be ensured only where groups of local stakeholders in rural areas agree to adopt a coordinated approach to resource management (Hodge 2001).

This highlights the need for a different approach regarding innovative forms of intervention, which takes into consideration the collective dimension of agri-environmental action in order to favour joint action to provide public goods at territorial scale. This kind of action, in fact, can provide a higher level of public goods provision which otherwise would not be possible to reach by individual farmers, and this is particularly relevant where the environmental objectives are related to the improvement of public goods such as biodiversity, landscape and water quality (ENRD 2010).

As will be further discussed in the next chapter, an increasing awareness has also been observed at the EU level recently, demonstrating that a more effective implementation of agri-environmental strategies could be secured by adopting territorial approaches to delivery, thus ensuring that the focus of action widens beyond the individual farm, in order to achieve sustainable solutions at landscape scale (Hart et al. 2011).

1.4 A New Model for Public Goods Provision in Rural Areas

It may be argued that the current model of public goods provision through the CAP described above is mainly based on the neo-classical definition of public goods (based on the characteristics of non-excludability and non-rivalry) and, as a consequence of this approach, the need for public intervention is defined as a result of market failures.

In the most comprehensive study on the CAP and the public goods provision financed by the European Commission (Cooper et al. 2009), as well as the in the proposals on the future CAP (European Commission 2011a; 2011b), a strong predominance of the positive approach to multifunctionality may be observed with a focus on the policy tools to be implemented rather than on the social and institutional innovations necessary to implement the public goods agenda.

Figure 1.3 schematizes this model, by showing the modalities of public intervention following the neo-classical assessment of public goods.

By taking into account the extended definition of public goods proposed here, based on the institutional economics theory, it is possible to recognize the main limitations of this model:

- The assessment of the publicness of goods according to the excludability/rivalry criteria does not take into consideration the public domain of the goods; namely the social construction of the problems concerned and the decision making process related to the strategies to be implemented;
- A narrow focus on policy tools aiming at modifying the land management practices at farm level, leaving behind the territorial and collective dimensions of the strategies to be implemented;

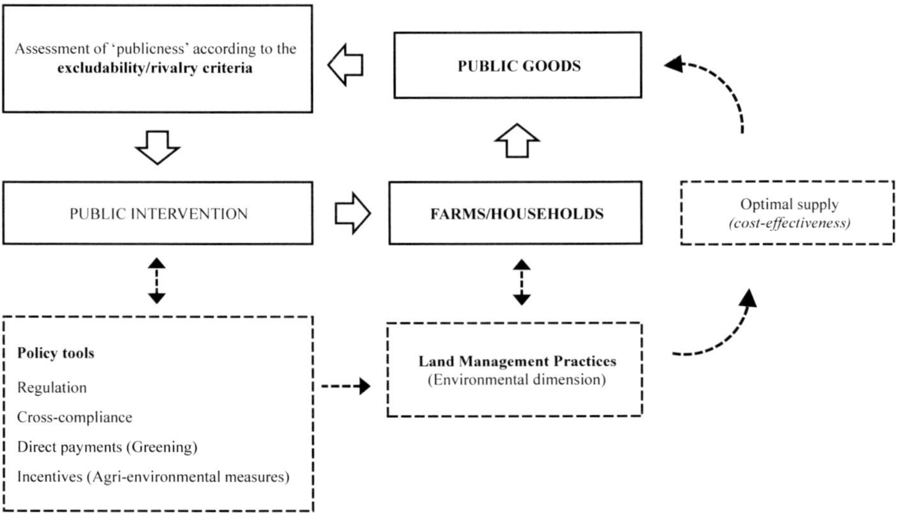

Fig. 1.3 The current model of public goods provision through the CAP

- An unsatisfactory recognition of the technical and institutional innovation needed in order to implement effectively the policy tools proposed;
- An unsatisfactory recognition of the knowledge and competencies needed to effectively implement the policy tools proposed;
- A narrow focus on the cost effectiveness of the measures proposed rather than on the effective level of public goods to be provided;
- A strong focus on the environmental dimension of the agricultural activities and on the environmental public goods (implicit to this model) (Fig. 1.3).

The conceptual limitations surrounding the conventional approach of public goods provision also highlight the need for a different approach regarding the forms of intervention, the institutional arrangements and the governance patterns to be implemented.

Indeed, in many cases the support of agriculture as a multifunctional activity has been translated into an indiscriminate support to the sector, since agriculture was considered *per se* a multifunctional activity. By contrast, the implementation of policies which are effective in public goods provision should involve a radical change in the way they are defined and implemented. Decoupled policies, such as the current CAP, which by definition are not linked to production, should have as their first objective the achievement of social and environmental objectives as well as the provision of immaterial goods, and their effectiveness and efficiency should be mainly evaluated in relation to these goals.

As discussed in sect. 1.1, an institutional approach to multifunctional agriculture provision broadens the scope of the analysis, by focusing, on one side, on a different classification of public goods associated with the primary sector and on, the other

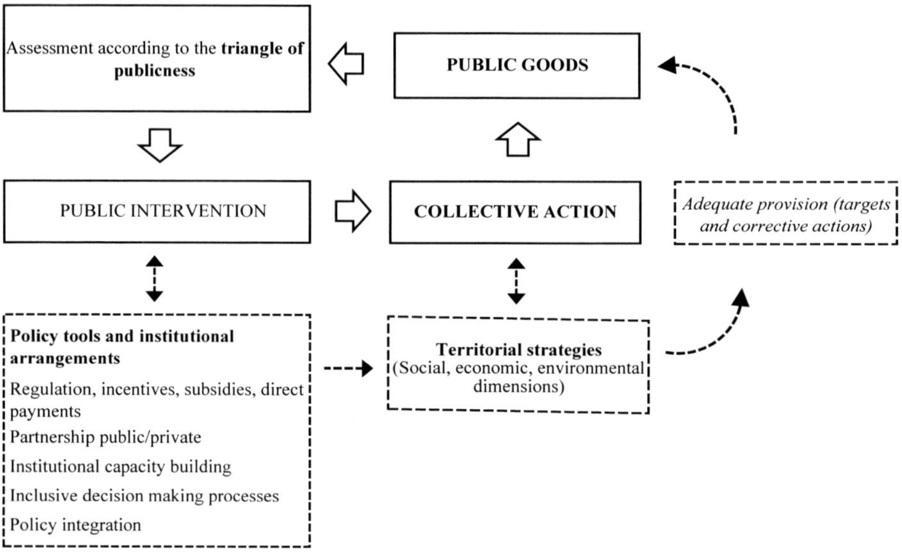

Fig. 1.4 A new model for public goods provision in rural areas

side, on the institutional arrangements and the territorial strategies which are aimed at providing adequate provision of such goods in rural areas.

This model highlights the need for focusing not only on the policy tools to be implemented, but also on the social and institutional arrangements and on the collective dimensions of the decision-making processes.

In Fig. 1.4 it is proposed that a different framework be used for public goods provision in rural areas, by showing the most significant implications that this new conceptualization could have for the definition and implementation of agricultural and rural development policies in the EU. Indeed, the characterization of goods according to the triangle of *publicness* allows for the consideration of other important dimensions, including the participatory nature of the process to place goods in the public domain and the distribution of benefits amongst the final users.

This model highlights the need for not focusing only on the policy tools to be implemented, but also on the social and institutional arrangements and on the collective dimensions of the decision-making processes.

The main differences between the conventional approach and this approach may be synthesized as follows:

- An extended role of public intervention, whereby policy tools are conceived together with the necessary social and institutional arrangements to make them effective;
- A better recognition of technical and institutional innovations as well as the knowledge and competencies needed to implement effectively the policy tools proposed;

- A focus on collective actions and local strategies rather than on single farms/households in order to increase the public goods provision in rural areas;
- A shift from the cost effectiveness of the measures to the adequate provision of public goods, by setting targets and by defining corrective actions;
- A broader vision of multifunctional agriculture which goes beyond the provision of (pure) environmental goods, but which aims at integrating the three dimensions (economic, environmental and social) of sustainability at local level.

The model of public goods provision described above calls for a more holistic interpretation of the concept of multifunctionality, where public intervention is not only deemed to regulate the supply and demand of public goods, but to create the governance structures which determine an adequate provision of public goods.

As will be further discussed in the next chapter, the provision of public goods is usually the result of the interactions of several stakeholders as well as the result of a deep change in farmers' practices which they undertake as result of collective strategies and joint learning.

Thus, the public intervention should be focused on this collective dimension, by supporting the collective action that is already in place and, in the other cases, by incentivizing the development of collectives and territorial strategies for public goods provision.

As it will be further discussed, in order to achieve these objectives it is necessary to implement effective policy tools, tailored and targeted to the local conditions, but also to adopt innovative institutional arrangements, public-private partnerships and other mixed solutions (i.e. co-operatives) which have the potential of ensuring effective strategies, tailored to the local needs.

In the following chapters the role of collective action as a viable alternative for providing environmental goods in rural areas is explored in more detail, by focusing on specific examples and on the related policy and institutional innovation. Indeed, few studies on the role of collective action for the provision of environmental public goods have been carried out in the EU context, and it is not clear to what extent collective action implemented specifically for the provision and/or protection of environmental goods could be supported and incentivized by the instruments traditionally adopted in the framework of the Common Agricultural Policy (CAP). At the same time, as argued by Ayer (1997), this does not imply that grass roots collective action always provides a more efficient (or better distributed) allocation of environmental benefits. Nevertheless, grass roots collective actions, which are little considered in the current debate, may represent interesting alternatives that are worth investigating further.

References

Anania et al (2009) A common agricultural policy for European public goods, 2009 declaration by a group of leading agricultural economists. http://www.reformthecap.eu/posts/declaration-on-cap-reform. Accessed 23 Feb 2011

Aumand A, Barthelemy D, Caron P (2006) Definitions, references and interpretations of the concept of multifunctionality in France. European Series on Multifunctionality 10:5–39

Ayer HW (1997) Grass roots collective action: agricultural opportunities. J Agr Resource Econ 22(1):1–11
Birdlife (2009) Proposal for a new EU common agricultural policy. http://cap2020.ieep.eu/vision/NGO-CAP-proposal.pdf. Accessed 12 Jan 2011
Brady M (2011) Impact of CAP reform on the environment: some regional results. In: OECD (ed) Disaggregated impacts of CAP reforms. OECD Publishing, Paris. http://dx.doi.org/10.1787/9789264097070-en
Conceição P (2003) Assessing the provision status of global public goods. In: Kaul P, Conceição K, LeGoulven, Mendoza RU (eds) Providing global public goods. UNDP, New York
Cooper T, Hart K, Baldock D (2009) The provision of public goods through agriculture in the European union. Institute for European Environmental Policy, London
Copa-Cogeca (2010) Il futuro della PAC dopo il 2013. http://www.copa-cogeca.be/Main.aspx?page=HomePage. Accessed 19 Dec 2010
Desjeux Y, Dupraz P, Thomas A (2011) Les biens publics en agriculture, une voie vers l'écologisation de la PAC. Colloque Ecologisation des politiques publiques et des pratiques agricoles. Avignon, 16–18 Mars
ENRD (2010) Conceptual framework on public goods provided through agriculture in the EU. http://enrd.ec.europa.eu/app_templates/filedownload.cfm?id=1B94F5A8-BD60-27F9-523E-AD522F83613D. Accessed 13 Jan 2011
European Commission (2005) Agri-environment measures overview on general principles, types of measures, and application. Directorate General for Agriculture and Rural Development, Unit G-4– Evaluation of Measures applied to Agriculture Studies
European Commission (2010) Communication from the commission to the European parliament, the council, the European economic and social committee and the committee of the regions—The CAP towards 2020: meeting the food natural resource and territorial challenges of the future, COM(2010)672 Final
European Commission (2011a) Proposal for a regulation of the European parliament and of the council establishing rules for direct payments to farmers under support schemes within the framework of the common agricultural policy. Brussels, 19.10.2011 COM(2011)625 final/2 2011/0280 (COD)
European Commission (2011b) Proposal for a regulation of the European parliament and of the council on support for rural development by the European agricultural fund for rural development (EAFRD). Brussels, 19.10.2011 COM(2011)627 final/2 2011/0282 (COD)
European Court of Auditors (2008) Is cross compliance an effective policy? Special report No. 8. Office for Official Publications of the European Communities, Luxembourg
Hagedorn K (2004) Multifunctional agriculture: an institutional interpretation. 90th EAAE Seminar on Multifunctional Agriculture, Policies and Markets: understanding the critical linkage. Rennes, 28–29 October
Hagedorn K, Artz K, Peters U (2002) Institutional arrangements for environmental co-operatives: a conceptual framework. In: Hagedorn K (ed) Environmental cooperation and institutional change: theories and policies for European agriculture. Elgar, Cheltenam
Hagedorn K (2008) Particular requirements for institutional analysis in nature-related sectors. Europ Rev Agr Econ 35(3):357–384
Hardin G (1968) The tragedy of the commons. Science 162:1243–1248
Hart K, Baldock D, Weingarten P, Povellato A, Pirzio-Biroli C, Osterburg B, Vanni F, Boyes A (2011) What tools for the European agriculture to encourage the provision of public goods. Study for the European Parliament's Committee on Agriculture and Rural Development
Hodge I (2001) Beyond agri-environmental policy: towards an alternative model of rural environmental governance. Land Use Pol 18(2):99–111
Kaul I, Conceição P, LeGoulven K, Mendoza RU (2003) Providing global public goods. UNDP, New York
Kaul I, Mendoza RU (2003) Advancing the concept of public goods. In: Kaul I, Conceição P, LeGoulven K, Mendoza RU (eds) Providing global public goods. UNDP, New York

Merlo M, Milocco E, Panting R, Virgilietti P (1999) La creazione di mercati per i beni e servizi ricreativo-ambientali collegati ad agricoltura e foreste: un'indagine in Austria, Germania, Italia ed Olanda. Rivista di EconomiaAgraria LIV(4):517–543

ODS (2002) Profiling the provision status of global public goods. ODS Staff Paper, Office of Development Studies. UNDP, New York

OECD (2001) Multifunctionality: towards an analytical framework. OECD Publishing, Paris

OECD (2006) The new rural paradigm: policies and governance. OECD Publishing, Paris

OECD (2008) Multifunctionality in agriculture: evaluating the degree of jointness, policy implications. OECD Publishing, Paris

OECD (2010) Environmental cross compliance in agriculture. OECD Publishing, Paris

OECD (2011) Evaluation of agricultural policy reforms in the European union. OECD Publishing, Paris

OECD (2012) Evaluation of agri-environmental policies: selected methodological issues and case studies. OECD Publishing, Paris. http://dx.doi.org/10.1787/9789264179332-en

OECD (2012) Farmer behaviour, agricultural management and climate change. OECD Publishing, Paris

Ostrom E (1990) Governing the commons: the evolution of institutions for collective action. Cambridge University Press, New York

Ploeg JD van der (2000) Revitalizing agriculture: farming economically as starting ground for rural development. Sociologia Ruralis 40(4):497–510

Ploeg JD, van der, Roep D (2003) Multifunctionality and rural development: the actual situation in Europe. In: van Huylenbroeck G, Durand G (eds) Multifunctional agriculture. A new paradigm for European agriculture and rural development. Ashgate, Hampshire

Prager K (2010) Applying the institutions of sustainability framework to the case of agricultural soil conservation. Environ Pol Govern 20(4):223–238

Randall A (2002) Valuing the outputs of multifunctional agriculture. Europ Rev Agr Econ 29(3): 289–307

Renting H, Rossing WA, Groot JC, Van der Ploeg JD, Laurent C, Perraud D, Stobbelaar DJ, Van Ittersum MK (2009) Exploring multifunctional agriculture. A review of conceptual approaches and prospects for an integrative transitional framework. J Environ Manage 90 (Suppl 2):S112-23. doi: 10.1016/j.jenvman.2008.11.014. Epub 2009 Jan 3.

RISE Foundation (2009) Public goods from private land, rural investment support for Europe Foundation, Brussels

Samuelson PA (1954) The pure theory of public expenditure. Rev Econ Statist 36:387–389

Van Huylenbroeck G, Durand G (2003) Multifunctional agriculture. A new paradigm for European agriculture and rural development. Ashgate, Hampshire

Van Huylenbroeck G, Vandermeulen V, Mettepenningen E, Verspecht A (2007) Multifunctionality of agriculture: a review of definitions, evidence and instruments. Living Rev Landsc Res 1(3):1–43

Vereijken PH (2001) Multifunctionality: applying the OECD framework. A review of literature in the Netherlands. OECD Publishing, Paris

Westhoek H van Zeijts H, Witmer M, van der Berg M, Overmars K, van der Esch S, van der Bilt W (2012) Greening the CAP: an analysis of the effects of the European Commission's proposal for the common agricultural policy. PBL Note Environmental Assessment Agency, The Hague

Wilson GA (2007) Multifunctional agriculture: a transition theory perspective. CABI publishing, Oxford

Zahrnt V (2009) Public money for public goods: winners and losers from CAP reform. ECIPE working paper, n. 8

Chapter 2
The Role of Collective Action

Abstract This chapter provides an overview of the definitions, role and characteristics of collective action, with a particular focus on the main strategies and dynamics regarding its role in proving public goods through agriculture. It is argued that a very relevant issue to be considered when analyzing the dynamics of collective action is what type of organization develops and manages such action, and this chapter provides a detailed description of the main institutional arrangements that may favour the development of grass roots collective action in rural areas. The first part of the chapter shows that some innovative institutional arrangements based on mixed private-public solutions, such as *co-production* and *co-management*, may represent effective territorial strategies to promote and support collective action related to the management of natural resources. The second part of the chapter describes the relevance of social capital in implementing those institutional arrangements in rural areas as well as the necessary shift towards a new structure of the agricultural knowledge and innovation systems in agriculture. The Agricultural Innovation System (AIS) should be re-configured in order to favor the implementation of collective strategies aimed at providing public goods through agriculture.

Keywords Collective action · Co-management · Co-production · Social capital · Agricultural Innovation System

2.1 Collective Action: Definition and Characteristics

During the last few decades an increasing amount of literature on collective action and natural resources has emerged, with a great emphasis on the conceptualization of collective action and on the analytical framework necessary to study it (Olson 1965; Wade 1987; Ostrom 1990).

Marshall (1998) defined collective action as 'the action taken by a group (either directly or on its behalf through an organization) in pursuit of members' perceived shared interests'. As observed by Meinzen-Dick et al. (2004), the more specific and varied definitions which have been added later have in common the following features: the involvement of a group of people, shared interests, common and voluntary actions to pursue those shared interests.

F. Vanni, *Agriculture and Public Goods*,

DOI 10.1007/978-94-007-7457-5_2,

A very relevant issue to consider when analyzing the dynamics of collective action is what type of organization has developed and/or supported such action. In many cases the outcomes of the collective action are highly dependent on the type of organizations involved, but also to the institutional arrangements which are in place at the local level. In the agricultural realm, for example, it is necessary to distinguish whether the collective action is developed by an organization directly controlled by farmers or if it is controlled and supported by a national/regional governmental authority.

From this perspective, Davies et al. (2004) distinguish two types of collective action: (i) *cooperation*: bottom-up, farmer-to-farmer collective action and (ii) *coordination*: top-down, agency-led collective action. While some bottom-up collective actions may receive government support, others may be carried out without government support. Similarly, some top-down collective actions are promoted by government policies but do not receive any support, while other collective actions receive support by local and/or government (OECD 2013).

This categorization implies the involvement of different levels of government (either central or local), which may provide the most effective support to the different strategies. From this perspective, the literature on collective actions and institutional arrangements for managing common pool resources has increasingly recognized the dynamic dimensions of institutions, which are context dependent and evolve over time. It is therefore necessary to understand how individuals interpret and respond to the different institutional arrangements in different contexts.

More specifically, a very important challenge for the analysis of collective action refers to the understanding of the role of formal and informal organizations that coordinate and support such actions, since in some cases these organizations exist only on paper and collective action occurs spontaneously, while in other cases institutions may play a vital role in creating and coordinating local action for a shared interest (Meinzen-Dick et al. 2004).

Although collective action is often associated with activities carried out by formal organizations, according to Ostrom (2004), more attention should be paid to informal collective action, where local networks or local groups of people organize and coordinate local action in order to achieve specific short-term purposes.

Since institutions play a crucial role regarding the development and the success of collective action, in the context of this book, particular attention is given to the implications of the institutionalizations of the more spontaneous and bottom-up collective actions. Indeed, as emphasized by Meinzen-Dick et al. (2004), any kind of collective action for routine maintenance will likely become institutionalized or integrated into mainstream policy frameworks (in the case of the EU agricultural sector mainly into the CAP) and while this institutionalization has the potential of reducing the transaction costs of negotiation, on the other hand the more institutionalized collective action, the less adaptable and flexible become.

The same authors argue that all the factors which influence the structure of groups and their organizations are relevant because they influence their conduct and then their outcomes. Indeed, neither the institutions involved nor the collective action itself are the ultimate objective; performance outcomes are important as well. As shown in

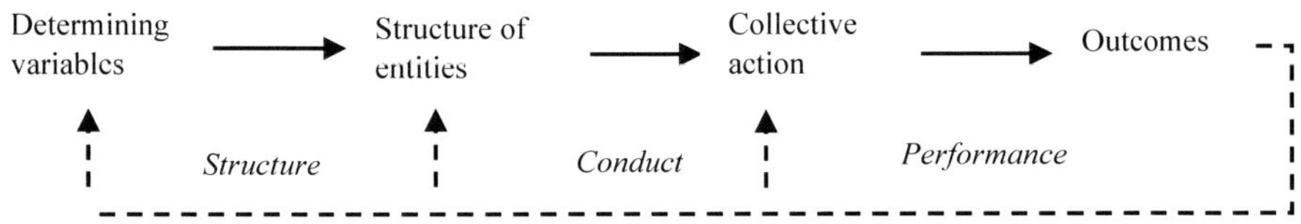

Fig. 2.1 Structure, conduct and performance of collective action. (Source: Meinzen-Dick et al. (2004))

Fig. 2.1, the many variables of interests present in collective actions usually determine feedbacks and co-movements that are likely to be very influential in determining the performance of collective action.

These feedbacks and co-movements are due to the fact that usually collective action is a dynamic process and, for this reason, it is also very difficult to measure directly, also because its performance relates to institutional settings and social relationships and it may vary over time, cultures and communities.

In order to analyze the performance of collective action related to natural resources and, more generally, to public and collective goods, some scholars have identified the substantive factors that it is necessary to take into account (Wade 1988; Ostrom 1990; Baland and Platteau 1996).

These factors have been grouped by Agrawal (2001) in a set of four basic categories:

1. Resource system characteristics;
2. Group characteristics;
3. Institutional arrangements;
4. External environment.

This conceptual framework, which has been used also by Davies et al. (2004) and by OECD (2013), allows the exploration of the key factors for successful collective action, as well as the barriers to be addressed in order to produce larger benefits through collective and territorial strategies.

According to this framework, collective action is highly affected by (1) the characteristics of the natural resources (type of goods) involved and on the knowledge and predictability of such resources, since information and communication on natural resources, for example related to technical requirements, are crucial issues to implement successful collective action. The type of knowledge usually includes both local knowledge and scientific expertise, and a successful integration of these two types of knowledge in many cases is a key issue to enable communities to use natural resources in a sustainable way (Agrawal 2001; Pretty 2003).

This factor is also related to (2) the characteristics of the group involved, which should have an appropriate size and homogeneity and, above all, it should allow the participants involved in the collective actions to increase their social relationships. These relationships, which may be synthesized through the concept of 'social capital', include trust, norms, reciprocity, obligations and expectations, values and attitudes, culture, information and knowledge, formal groups/associations, institutions, rules and sanctions (Davies et al. 2004).

The success of collective action is also determined by (3) the involved institutional arrangements which, according to the main studies on the topic (Ostrom 1990; Wade 1988), should involve on one side, locally devised and simple rules and, on the other side, they should rely on effective monitoring and sanction systems. In addition to the rules implemented, the success of local strategies is also linked to the 'thickness' of local institutions, which may be able to generate public objectives from economic activities. 'Institutional thickness' in a given territory is linked to the combination of 'human capital' (knowledge resources), 'social capital' (trust, reciprocity and other social relations) and 'political capital' (capacity for collective action) (Mantino 2010).

Finally, external forces and authorities (4) also affect collective action to a large extent, and these forces may be interpreted as both financial and non-financial support. Financial support is particularly relevant at the initial stage of the collective action, since it usually involves higher transaction costs compared to individual activities (Mills et al. 2010). Conversely, non-financial support is related to the need of governments to play a pro-active role in setting basic rights, guidelines, rules (also with penalties and sanctions) and public objectives which may encourage collective action (Ayer 1997).

2.2 Barriers and Benefits of Collective Action

The role of collective action is increasingly analyzed also in the context of agriculture and rural development. The majority of studies and analyses are related to collective marketing initiatives, since a collective and coordinated approach of farmers in the food supply chain may have positive economic effects, by increasing the economies of scale and by reducing transaction costs.

At the same time, as it will be further discussed in the following chapters, it is increasingly recognized that the collective action of farmers and of other rural stakeholders may also play an important role in delivering public goods, non-commodity outputs and environmental services (Polman et al. 2010).

In the environmental realm, a joint action can be undertaken by farmers' organizations, farmers' associations or by an informal group of farmers for many reasons, such as reducing the transaction costs to collect information on innovative (and more sustainable) production practices, to comply with new legislation, to take market opportunities (i.e. to negotiate a premium price with the large distribution channels) or to monitor the jointness between commodity and non-commodity output (Van Huylenbroeck 2008).

As demonstrated by Uetake (2012) in the case studies from New Zealand, in many cases the provision of agri-environmental public goods through collective action also allows obtaining larger benefits, which were identified in the following areas: scale merits, sharing knowledge and increasing capacity and tackling local issues.

Nevertheless, successful initiatives must overcome the most common barriers to collective action, which are traditionally identified in the problem of free riding and in higher transaction costs.

With regard to the benefits of collective action, the issue of scale is particularly important, since collective action may have ecological scale merits and may improve the economy of scale and scope. As shown by some studies (Davies et al. 2004; Mills et al. 2010), since environmental public goods such as biodiversity and landscape in many cases cannot be provided by single farmers, collective action allows the addressing of the problem of public goods provision at the geographically and ecologically appropriate scale. In addition, by mobilizing territorial resources in a coordinated way, collective action may reduce the costs of public goods provision (economy of scale) and may improve the co-ordination mechanisms for the joint provision of several public goods (economy of scope).

As will be further discussed in Sect. 2.5, another of the key benefits of collective action is the possibility of sharing knowledge and learning for the stakeholders who take place in the collective initiatives. In many cases a cooperation approach relies heavily on the local knowledge of stakeholders and on the possibilities to integrate this knowledge into the decision making process. Thus, collective action increases the credibility and legitimacy of decision-making, but also allows the collecting and sharing of information at lower costs compared to the individual approaches.

The other important benefit of collective action is the possibility of tackling efficiently local issues. In many cases central governments have increasing difficulties in tackling local issues and cannot find a viable solution for local problems, while through collective action it is possible to implement strategies tailored to local problems, since it allows greater flexibility, responsiveness and local relevance (Davies et al. 2004). As will be further discussed in the case studies described in the book, in many cases civil society associations are the typical initiators of innovative development steps and their involvement usually contributes to a great extent to the success of such initiatives.

Together with the aforementioned benefits, the literature acknowledges that free riding and higher transaction costs may represent important barriers to collective action.

Free riding as one of the main barriers to collective action was already identified in the seminal work of Olson (1965), who showed the way in which collective action often involves some individuals who tend not to contribute to group activities because they benefit from other member's activities. This problem is particularly relevant when collective action takes place with the objective of public goods provision, since the benefits of public goods, which are non-rival and non-excludable, cannot be limited to the group members of collective action. At the same time, as it will be demonstrated in the following chapters, in grass roots collective actions traditional incentive mechanisms such as trust, solidarity reciprocity and reputation may be determinant factors in the reduction of free riding.

With regard to the transaction cost problem, the literature clearly acknowledges that, especially in the initial phase of their implementation, collective initiatives may have higher transaction costs compared to individual actions (Ostrom 1990;

Table 2.1 Transaction costs in collective action. (Source: Singleton and Taylor (1992))

Transaction costs	Examples
Search costs incurred in identifying possibilities for mutual gains	Identification of funding sources
	Identification of relevant stakeholders
	Cost of gathering information
Bargaining costs associated with negotiating an agreement	Time spent at meetings
	Effort expended in verbal and written communications
Monitoring and enforcement costs involved in making sure all parties keep to the agreement	Employment of external monitor
	Time and effort spent monitoring others informally

Davies et al. 2004). This has important implications when implementing policies to support collective strategies for public goods, since the collective action usually involves higher costs related to search costs (incurred in the identification of possibilities for mutual gains), bargaining costs (associated with negotiation and agreement) and monitoring and enforcement costs (Singleton and Taylor 1992; Table 2.1).

However, while some additional costs are inevitable in order to make collective action works, in many cases the benefits of collective action cover these costs (Uetake 2012) because of the economy of scale and scope (Hodge and McNally 2000; Davies et al. 2004). Finally, as will be further discussed in the following sections, the benefits of collective action may be maximized when appropriate institutional arrangements are in place and where the role of social capital is enhanced as result of the effects of social networks, trust and reciprocity among group members.

2.3 Institutional Arrangements

A very relevant issue to address when assessing the dynamics of collective action in the context of agriculture is what type of organization is involved, since the dynamics related to the drivers, processes and outcome may differ to a large extent; for example, between an organization controlled directly by farmers or controlled by a national/regional government.

In the context of western countries, especially in the European Union, the public intervention in agriculture is quite centralized, and central governments still play a very crucial role. Although during the last decades important efforts to increase the decentralization and to promote local defined rural and agricultural policies were observed, it is evident that, especially at the EU level, the agriculture is a highly subsidized and the economic performance of the sector is highly dependent on public support, that is usually implemented through rather top-down policy tools (see sect. 1.3).

Conversely, the concept of collective action itself suggests the need to look beyond the simple top-down management—that in the agricultural sector is usually based on state intervention—but also to look at the public/private partnerships and those innovative institutional arrangements which involve different levels. Those multi-stakeholder arrangements are usually characterized by strong horizontal linkages among user groups at the same level of organization, but also by vertical linkages between different levels, for example between local stakeholders and central governmental agencies (Berkes 2009, p. 1693).

As will be further discussed in next few chapters, this multi-stakeholder approach is also very relevant to stimulate collective action related to provision and protection of public goods (increase of positive externalities or decrease of negative externalities). Indeed, the effectiveness of the strategies for the provision of public goods and services is usually related to the involvement of a broader range of rural stakeholders that act together by sharing different knowledge, expertise and commitment for common goals.

As highlighted by Gatzweiler (2006, p. 300), farmers cannot be expected to be the sole carrier of the costs for providing public goods and services and the government cannot be the sole authority in the allocation of public funding, but in many cases it is necessary to seek ways towards mixed solutions. From this point of view, as discussed by many authors (Hagedorn et al. 2002; Van Huylenbroeck et al. 2009), the solution is not as easy as leaving the allocation problem of private goods to the market and that of public goods to the government, but it is usually necessary to explore innovative solutions, based on mixed public-private arrangements which could ensure an effective provision of public goods through collective and inclusive strategies. Thus, it may be argued that collective action for public goods through agriculture does not involve merely larger areas owned by many farmers, but also innovative institutional arrangements and coordinating mechanisms implemented at the right scale. For example, in the case of public footpaths created for connecting across individual farms, landholders need to co-ordinate decisions in order to create networks of paths that can offer a worthwhile recreation experience. Thus, in many cases it is necessary to overcome the traditional environmental contracts represented by a direct link between an agency and a land user. Other arrangements may be more effective, by revealing demand within a market context, by establishing incentives for landholders to co-ordinate their actions and by reducing the requirements for public expenditure (Hodge 2001).

Some innovative institutional arrangements that may represent the basis for collective action regarding the provision of agri-environmental public goods have been conceptualized through the definitions of *co-management* and *co-production*.

Co-management, defined as the sharing of power and responsibility between the government and the local resource users is a hybrid regime combining centralized and decentralized state and community institutions (Berkes 2009, p. 1692). More particularly, co-management was defined by Singleton (1998, p. 7) as a 'governance system ... that combine[s] state control with local, decentralized decision-making and accountability which, ideally, combine[s] the strengths and mitigate[s] the weakness of each'. This institutional arrangement is usually combined with learning-based approaches, since it may be considered a knowledge partnership where different levels of organizations have comparative advantages in generating and mobilizing the knowledge acquired at different scales.

As described by Singleton (2002, p. 3), the appeal of co-management is related to both efficiency and legitimacy. The efficiency is related to the availability of higher quality and less costly information, since in co-management arrangements local knowledge is usually combined with scientific knowledge produced by state agency scientists. The integration of these two types of knowledge may result in producing a more complete, finely-tuned set of information. At the same time, monitoring and enforcement can be more effective by virtue of being carried out by local people involved in the collective action. Similarly, the legitimacy of the system is enhanced by the fact that user-groups and community members are involved, which may result in people being more willing to comply voluntarily with the specific requirements in place.

While co-management refers to an arrangement in which private organizations or associations produce services in collaboration with the state, *co-production* refers to an arrangement where, at least in part, citizens produce their own services. Co-production has been defined by Ostrom (1996, p. 1073) as 'the process through which inputs used to produce a good or services are contributed [to] by individuals who are not "in" the same organization [. . .]. Co-production implies that citizens can play an active role in producing public goods and services of consequence to them'.

The co-production concept, analyzed mainly by American scholars in public administration studies, was born out of an acknowledgement that production of a service, as contrasted to a good, was difficult without the active participation of those who are supposed to receive the service. Thus, the term co-production describes the potential synergy and collaboration that could occur between the provider of services (usually the government) and the users of services (usually the citizens), by showing different and possible roles of individuals or groups in the production of such services.

As shown by Ostrom (1996), reciprocity is an important requisite to make co-production advantageous. Indeed the co-production process implies the building of credible commitment of the participants to one another and clear and enforceable contracts between government agencies and citizens which enhance that credibility.

The added values of these decentralized and hybrid regimes is due to the fact that when implementing a co-management of co-production approach, different actors need to work and think together, and deliberate to generate new knowledge or make sense of knowledge from different sources (Berkes 2009, p. 1695). As argued by Davidson-Hunt and O'Flaherty (2007, p. 293): 'working from the premise that knowledge is a dynamic process—that knowledge is contingent upon being formed, validated and adapted to changing circumstances—opens up the possibility . . . to establish relationships with indigenous peoples as co-producers of locally relevant knowledge'.

Finally, it may be argued that institutional arrangements based on co-production and on co-management imply a shift from a linear approach to policy design towards a policy cycle where rural development strategies are designed and implemented according to local needs, where the knowledge of local communities play a pivotal role. Indeed, the conventional approach to public goods provision usually assumes a greater amount of knowledge within the government than is actually available and, by failing to acknowledge society's learning abilities, usually makes insufficient use of social dynamics for realizing public objectives (Hajer 2011, p. 26). On the opposite side, an approach based on co-management and co-production implies a renewed role for government, which should favour experiments, innovation and learning processes, since government is responsible for setting public objectives but there is also a need for an increasing role of society, which is the carrier of required change. At the same time, as pointed out by Enengel et al. (2011, p. 1266) these innovative institutional arrangements are processes of learning by doing from previous and ongoing experiences, and must be critically analyzed in order to derive recommendations for future practices.

2.4 The Role of Social Capital

Many studies have pointed out the importance of social capital for collective action (Pretty 2003; Rudd 2000; Ostrom 2000).

Social capital has been defined by Bourdieu (1986, p. 251) as 'the sum of the resources, actual or virtual, that accrue to an individual or a group by virtue of possessing a durable network of more or less institutionalized relationships of mutual acquaintance and recognition'. Similarly, Putnam et al. (1993) define social capital as 'features of social organizations, such as networks, norms and trust that facilitate coordination and cooperation for mutual benefits'. The other definitions which have been added later have in common a strong focus on the 'relations' and on the benefits, such as mutual cooperation or various other resources, that result from these relations (Dahal and Adhikari 2008).

As argued by Dahal and Adhikari (2008, p. 3), the studies on collective action have widely used the definition of Putnam et al. (1993), since: (i) social capital in collective action is usually related to meso and collective units, such as associations, communities and regions; (ii) social capital is presented as a solution of the barriers of collective action and (iii) the social capital framework is applied to the study of the performance of institutions, such as regional governments.

An approach based on social capital and social norms may be effectively applied to the collective management of environmental resources, since it can complement the traditional public policy approaches based on regulation, taxation and pricing to address environmental problems (World Bank 2009).

In analyzing the role of social capital and the collective management of resources, Pretty (2003, p. 1913) emphasizes the importance of the following four features: (i) relation of trust; (ii) reciprocity and exchanges; (iii) common rules, norms and sanctions and (iv) connectedness in networks and groups.

Relations of trust (i) are an important prerequisite to work-cooperatively, and this is particularly true in the case of collective action, where mutual trust plays

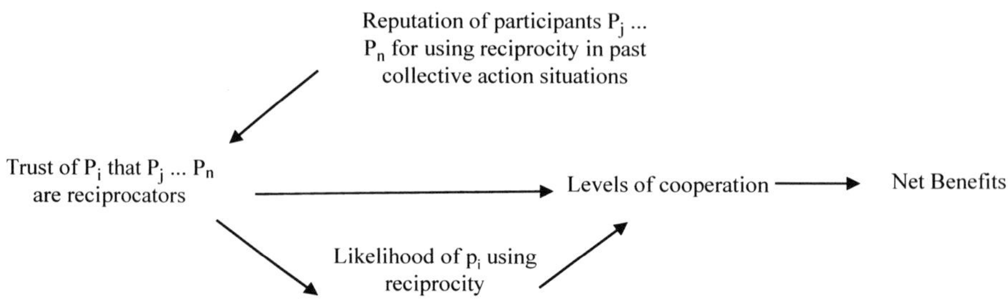

Fig. 2.2 Relationships between trust, reputation, reciprocity and collective action. (Source: Ostrom (2007))

a central role in reducing transaction costs between people, by avoiding the need to monitor others and thereby save money. Nevertheless, as argued by Baland and Platteu (1996), trust is easier to establish in societies and organizations with long and established traditions of co-operation.

Trust is strongly related to reputation. Indeed, when paying attention to reputation, people are more bounded by mutual obligation and reciprocity and, as observed by Wade (1988), in many cases collective action increases the chances to be successful when people are concerned about their social reputation.

Similarly, Ostrom (2007) has observed that participants involved in collective action in many cases decide to trust other participants on the basis of their reputation in past collective action situations. According to Ostrom, at the core of successful collective action are the links between the trust that one participant (Pi) has in the others (Pi, ..., Pn) involved in a collective action situation, the investment others make in trustworthy reputations, and the probability of all participants using reciprocity norms. As shown in Fig. 2.2, the joint action of reputation, trust and reciprocity is very important, since these factors positively reinforce each other and this results in increasing cooperation and, as a consequence, in increasing the net benefits of collective action.

Cooperation may also be promoted by (ii) reciprocity and exchange, since reciprocity increases trust and contributes to the development of long-term obligations between people, which helps in achieving positive environmental outcomes (Pretty 2003). This reciprocity must be based on trust, since trust 'lubricates co-operation' between people and, by reducing transaction costs, may liberate important financial resources (Pretty and Ward 2001).

Moreover, successful collective action is also dependent on (iii) common rules, norms, and sanctions which must be set up according to an inclusive criterion in order to ensure that group interests are in line with those of individuals and should be effective in changing behaviours. These 'rules of the game' should be also proposed and recognized as much as possible by people participating in collective action, in order to be effective in giving individuals the confidence to invest in the collective goods.

Finally, with regard to the (iv) connectedness in networks and groups, it is possible to recognize three types of connectedness: bonding, bridging and linking (Pretty 2003):

- *Bonding* describes the links from membership of groups of similar people with strong ties, such as networks of friends, family and associates;
- *Bridging* from membership of more diverse associations, with weaker links between individuals, such as interests groups or social and leisure clubs;
- *Linking* describes connections with people in position of power, and is good for accessing support from formal institutions.

In the context of this book, in analyzing the role of social capital for the success of collective action, particular attention is given to all these aspects of social capital regarding two types of interactions: the interactions within the farmers' communities and the interactions between farmers and the other local rural stakeholders.

Indeed, farmers' decision making processes are generally strongly influenced by the judgments of their peers, and this emphasizes the need to explore the individual interests which allow farmers to interact with each other, in order to understand the social networks, trust and norms of reciprocity which are in place in the farming communities under study. As shown by Carolan (2006), the acknowledgement of farmers' knowledge and the increasing interactions amongst peers may allow individuals to develop 'interactional expertise in bringing together knowledge produced in different contexts'.

At the same time, the study of collective action implies a territorial and integrated approach, where it is also important to analyze the interactions of farmers with wider networks, which may involve:

- Other farming communities, also involving forms of cooperation both at local and at extra-local levels;
- Institutions (local and regional administration, public bodies and independent institutions);
- Other organizations (NGOs, environmental organizations, political groups);
- Citizens and consumers (local inhabitants, tourists, consumers, cultural associations);
- Technical services and experts who play a role in increasing the spread of knowledge and innovation amongst farmers.

Even though there is increasing evidence that a high level of social capital is usually a central requisite for a collective effective provision/preservation of agri-environmental goods, it must be stressed that social capital 'is not easy to find, see and measure as is physical capital' (Ostrom 2000, p. 188).

Moreover, from a public goods perspective, the main challenges are translating social capital into norms or values that discourage behaviours which cause externalities, and by strengthening community ties so that sanctions may be provided against those who transgress (DEFRA 2008, p. 17).

As will be discussed in the next section, this important challenge is strongly related to the role of learning and innovation in agriculture. Advisory systems, extension, diffusion of innovation and training have a crucial role in shaping the attitudes and motivation of farmers and in determining important drivers for the success of collective action.

2.5 A New Approach to Learning and Innovation

When successful collective actions for public goods take place, the social dynamics enhance the capacities of farmers, since collective action makes it possible for participants to collect and share knowledge and information, also with other rural stakeholders and landholders that together utilize their knowledge, skills and institutions (Hodge and Reader 2007).

However, the need of favouring the joint production of knowledge and joint learning are very increasingly recognized, more generally, as one of the main challenges for the Agricultural Knowledge and Innovation Systems. In a recent EU report on the topic, for example, the authors recommend building on models of joint knowledge-production, spanning the boundaries between knowledge generators and users, since expertise is derived from multiple sources (academics, businesses, land managers and the society) and each actor can make a valuable contribution to the knowledge base (EU SCAR 2012).

The formal Agricultural Knowledge System (AKS), based on the triangle of agricultural research, education and extension, has been the dominant paradigm for agricultural extensions in the 1970s and 1980s. This system, rooted on the so called 'knowledge transfer paradigm', was based on the dissemination of information and technical solutions amongst farmers, with the objective of diffusing the adoption of predetermined practices.

This paradigm is based on a strong confidence in scientific progress, whereby the innovations are created by the scientific community and the new technologies are 'transferred' to farmers who 'adopt' them, through a top-down linear process, from research to farmers (Knickel et al. 2009).

This approach has been strongly criticized, and criticisms may be grouped under three main concerns: (i) it is not appropriate for the modern multifunctional agriculture; (ii) it does not reflect the empirical evidence of how farmers use information and (iii) it does not take into adequate consideration the multiple sources from which knowledge is generated, and the other influences of farmers, notably also by the information received by other farmers (Blackstock et al. 2010).

Here it is particularly relevant to highlight the limits of the traditional AKS model in addressing the provision of agri-environmental public goods through agriculture. Indeed, the mission of AKS was mainly increasing the productivity of agriculture, usually relatively to a narrow range of crops. Conversely, the changing scenario of the global production system, involving the diversification on the demand on agriculture, has shown in quite a clear way the incoherence of the dominant production system and the related accompanying AKS: an increasing gap between private and public goals was observed with a consequent emergency of different scientific and technological paradigms (Brunori et al. 2011).

For these reasons, during the 1990s and 2000s an alternative 'human development' approach on innovation was developed, which emphasizes the need for farmers to develop their own solutions to the problems, where the role of extensions is to facilitate interaction, learning and innovation rather than persuading farmers on the practices to be adopted. Thus, the agricultural innovation literature has developed the concept of Agriculture Innovation Systems (AIS) (Spielman and Birner 2008), which were defined as 'networks of organizations, enterprises, and individuals focused on bringing new products, new processes, and new forms of organization into economic use, together with the institutions and policies that affect the way different agents interact, share, access, exchange and use knowledge' (World Bank 2006).

In order to respond to the increasing complexity of the primary sector and to the need of increasing the provision of public goods, in some countries innovation processes previously based on strong hierarchical patterns, have been increasingly substituted by systems structured as networks. A network-like structure of innovation patterns gives more flexibility and allows for the primary sector to respond to the emerging consumers demand as well as the new policies implemented. Indeed, while the AKS uni-linear approach, by focusing on the persuasion of individuals, fails to explain the decision-making processes within specific social systems and the related collective behavioral changes, the new AIS paradigm emphasizes the process of networking and interactive learning among a heterogeneous set of actors, such as farmers, input industries, processors, traders, researchers, extensionists, government officials, and civil society organizations.

From this perspective, it may be argued that the provision of agri-environmental public goods through collective action is particularly challenging also from a knowledge and innovation perspective. Indeed, the new paradigm, based on AIS, in order to be effective in facilitating the collective provision of public goods, must create new spaces for social and institutional innovation, by linking together different disciplines, different administrative and policy spheres and encouraging new learning processes.

In addition, moving from an approach based on knowledge transfer' towards processes that facilitate 'knowledge exchanges' involves recognizing the crucial role of local and tacit knowledge, where an important feature is social innovation in multi-stakeholder collective practices and knowledge production (Schmid et al. 2012).

The central issue of new approaches on innovation (AIS) is how farmers, scientists and advisors can co-create, collectively, new meanings and new codes for sustainable practices. This process, which involves a transformation of values, norms, rules and powers amongst the actors, has been conceptualized as 'social learning' (Röling and Wagemakers 1998). Social learning is defined as 'learning that occurs when people engage one another, sharing diverse perspectives and experiences to develop a common framework of understanding and basis for joint action' (Schusler et al. 2003, p. 311). Thus, social learning may be considered an iterative process of knowledge co-production amongst stakeholders brought into interaction: when stakeholders become aware of how other stakeholders understand reality and how these 'understandings' relate to practical experiences and defended interests (Steyaert et al. 2007, p. 540).

As argued by Blackstock et al. (2010, p. 5637), within differentiated farming cultures there are different individual and group identities, which influence and interpret knowledge and the consequent behaviour changes in different ways. Therefore, understanding and influencing behaviour is a complex and multi-faceted issue, also related to issues of power and politics within farming cultures and between these groups and the wider policy and political settings. An understanding of relations within and outside the actors involved in the collective action, as well as the power dynamics, where different interests or influences are expressed, is crucial to fully understanding the process of knowledge exchanges and the collective decision making processes. Thus, it may be argued that the achievement of new environmental objectives for agriculture involves significant changes in farmer's practices and it is related to two dimensions: a technical change but also a more structural shift regarding their professional identity and their role in the management of the territory (Deuffic and Candau 2006, p. 565).

In order to provide successful technological solutions for sustainable agriculture, such as no-tillage, integrated pest management and organic farming, such solutions need to be integrated with broad cultural, social, political and economic transformations (Schneider et al. 2009). The new AIS approach must then involve a process of co-production of knowledge between academic and non-academic actors and above all, a process of knowledge co-production between farmers, experts and scientists.

As emphasized by Tarnoczi and Berkes (2010), information from government and producer organizations can be very important for the co-production of knowledge, and especially producer organizations may undertake bridge building, by linking policy makers to farmers, also by including local knowledge and local experiences when implementing extension strategies at the farm-level. Indeed the new AIS approach, based on the principles of empowerment and participation, emphasizes the key role of the non-expert form of knowledge and the role of extensions in facilitating collective processes. As highlighted in the EU SCAR position paper (2012, p. 9), innovation is a risky business and benefits from the exchange of ideas, learning and innovation networks have proven to be an adequate vehicle for empowering groups of farmers to investigate new options to make their business more viable and sustainable.

The case studies analyzed in this book (chap. 4 and 5) show how this alternative approach to learning and innovation, based on the AIS principles, has been successfully experimented in two grass roots collective actions, by showing that this innovative system, based on co-management, co-production of knowledge and social learning, may represent an effective approach for the provision of public goods through agriculture.

References

Agrawal A (2001) Common property institutions and sustainable governance of resources. World Devel 29(10):1649–1672
Ayer HW (1997) Grass roots collective action: agricultural opportunities. J Agric Resource Econ 22(1):1–11
Baland JM, Platteau JP (1996) Halting degradation of natural resources: is there a role for rural communities? Cambridge University Press, Cambridge
Berkes F (2009) Evolution of co-management: Role of knowledge generation, bridging organizations and social learning. J Environ Manage 90(5):1692–1702
Blackstock KL, Ingram J, Burton R, Brown KM, Slee B (2010) Understanding and influencing behaviour change by farmers to improve water quality. Sci Total Environ 408(23):5631–5638
Bourdieu P (1986) Forms of capital. In: Richards JC (ed) Handbook of theory and research for the sociology of education. Greenwood Press, New York
Brunori G, Berti G, Klerkx L, Tisenkopfs T, Roep D, Moschitz H, Home R (2011) Learning and Innovation Networks for Sustainable Agriculture. WP2—Conceptual Framework, SOLINSA EU project
Carolan MS (2006) Social change and the adoption and adaptation of knowledge claims: whose truth do you trust in regard to sustainable agriculture? Agric Human Values 23(3):325–339
Dahal GR, Adhikari KP (2008) Bridging, linking, and bonding social capital in collective action. Capri Working paper No. 79
Davidson-Hunt IJ, O' Flaherty RM (2007) Researchers, indigenous peoples and place-based learning communities. Soc Natur Resour 20(4):291–305
Davies B, Blackstock K, Brown K, Shannon P (2004) Challenges in creating local agri-environmental cooperation action amongst farmers and other stakeholders. The Macaulay Institute, Aberdeen
DEFRA (2008) Understanding Behaviours in a Farming Context: Bringing Theoretical and Applied Evidence together from Across Defra and Highlighting Policy Relevance and Implications for Future Research, November. Defra Agricultural Change and Environment Observatory Discussion Paper
Deuffic P, Candau J (2006) Farming and landscape management: how French farmers are coping with the ecologization of their activities. J Agric Environ Ethics 19(6):563–585
Enengel B, Penker M, Muhar A, Williams R (2011) Benefits, efforts and risks of participants in landscape co-management: an analytical framework. J Environ Manage 92(4):1256–1267
EU SCAR (2012) Agricultural Knowledge and Innovation Systems in Transition: a Reflection Paper. http://ec.europa.eu/research/agriculture/scar/pdf/akis_web.pdf. Accessed 13 Dec 2012
Gatzweiler F (2006) Organizing a public ecosystem service economy for sustaining biodiversity. Ecolog Econ 59(3):296–304
Hagedorn K, Artz K, Peters U (2002) Institutional arrangements for environmental co-operatives: a conceptual framework. In: Hagedorn K (ed) Environmental cooperation and institutional change: theories and policies for European agriculture. Elgar, Cheltenam
Hajer M (2011) The energetic society. In search of a governance philosophy for a clean economy. PBL Netherlands Environmental Assessment Agency, The Hague
Hodge I (2001) Beyond agri-environmental policy: towards an alternative model of rural environmental governance. Land Use Pol 18(2):99–111
Hodge I, McNally S (2000) Wetland restoration, collective action and the role of water management institutions. Ecolog Econ 35(1):107–118
Hodge I, Reader M (2007) Maximising the Provision of Public Goods from Future Agri-environment Schemes. Final Report for Scottish Natural Heritage, Rural Business Unit, Department of Land Economy, University of Cambridge
Knickel K, Brunori G, Rand S, Proost J (2009) Towards a better conceptual framework for innovation processes in agriculture and rural development: from linear models to systemic approaches. J Agric Edu Ext 15(2):131–146

Mantino F (2010) Understanding delivery mechanisms in EU rural development policies: an institutional approach. Paper presented at the 118th Seminar of the EAAE "Rural Development: governance, policy design and delivery", Ljubljana, Slovenia August 25–27
Marshall G (1998) A dictionary of sociology. Oxford University Press, New York
Meinzen-Dick R, Di Gregorio M, McCarthy N (2004) Methods for studying collective action in rural development. Agric Syst 82(3):197–214
Mills J, Gibbon D, Ingram J, Reed M, Short C, Dwyer J (2010) Collective action for effective environmental management and social learning in Wales. Proceedings of the 9th European IFSA Symposium, 4–7 July 2010. Vienna, Austria
OECD (2013) Providing Agri-environmental Public Goods through Collective Action. OECD Publishing, Paris http://dx.doi.org/10.1787/9789264197213-en
Olson M (1965) The Logic of Collective Action. Harvard University Press, Cambridge
Ostrom E (1990) Governing the commons: the evolution of institutions for collective action. Cambridge University Press, New York
Ostrom E (1996) Crossing the great divide: coproduction, synergy, and development. World Devel 24(6):1073–1087
Ostrom E (2000) Social capital: a fad or a fundamental concept? In: Dasgupta P, Sarageldin I (eds) Social capital. A multifaceted perspective. The World Bank, Washington
Ostrom E (2004) Understanding collective action. In: Meinzen-Dick R, Gregorio M Di (eds) Collective action and property rights for sustainable development, 2020 vision for food, agriculture and the environment. Focus 11, IFPRI International Food Policy Research Institute, Washington
Ostrom E (2007) Collective Action Theory. In: Boix C, Stokes S (eds) The Oxford handbook of comparative politics. Oxford University Press, Oxford
Polman N, Poppe KJ, Schans JW van Der, Ploeg JD van der (2010) Nested markets with common pool resources in multifunctional agriculture. Rivista di Economia Agraria LXV(2):295–318
Pretty J (2003) Social capital and the collective management of resources. Science 302:1912–1914
Pretty J, Ward H (2001) Social capital and the environment. World Devel 29(2):209–227
Putnam RD, Leonardi L, Nanetti R (1993) Making democracy work. Civil traditions in modern Italy. Princeton, New Jersey
Röling NG, Wagemakers MAE (eds) (1998) Facilitating sustainable agriculture: participatory learning and adaptive management in times of environmental uncertainty. Cambridge University Press, Cambridge
Rudd MA (2000) Live long and prosper: collective action, social capital and social vision. Ecolog Econ 34(1):131–144
Schmid O, Padel S, Levidow L (2012) The bio-economy concept and knowledge base in a public goods and farmer perspective. Bio base. Appl Econ 1(1):47–63
Schneider F, Fry P, Ledermann T, Rist S (2009) Social learning processes in Swiss soil protection—the 'from farmer—to farmer' project'. Hum Ecol 37(4):475–489
Schusler TM, Decker DJ, Pfeffer MJ (2003) Social learning for collabourative natural resource management. Soc Nat Resour 16(4):309–326
Singleton S (1998) Constructing cooperation: the evolution of institutions of comanagement. University of Michigan Press, Ann Arbor
Singleton S, Taylor M (1992) Common property, collective action and community. J Theor Polit 4(3):309–324
Singleton S (2002) Cooperation or capture? The paradox of co-management and community participation in natural resource management and environmental policy making. Environ Polit 9(2):1–21
Spielman DJ, Birner G (2008) How innovative is your agriculture? using innovation indicators and benchmarks to strengthen national agricultural innovation systems. Agriculture and Rural Development Discussion Paper 41. The World Bank, Washington DC
Steyaert P, Barzman M, Billaud JP, Brives H, Hubert B, Ollivier G, Roche B (2007) The role of knowledge and research in facilitating social learning among stakeholders in natural resources management in the French Atlantic coastal wetlands. Environ Sci Pol 10(6):537–550

Tarnoczi TJ, Berkes F (2010) Sources of Information for Farmers' Adaptation Practices in Canada's Priairie Agro-Ecosystem. Climatic Change 98(1–2):299–305

Uetake T (2012) Providing agri-environmental public goods through collective action: lessons from New Zealand case studies. Paper presented at the 2012 NZARES Conference, Tahuna Conference Centre—Nelson, New Zealand. August 30–31

Van Huylenbroeck G (2008) Market and rural policy institutions to stimulate multifunctional food and fibre production. In: Cesaro L, Gatto P, Pettenella D (eds) The Multifunctional Role of Forests—Policies, Methods and Case Studies. EFI Proceedings No. 55

Van Huylenbroeck G, Vuylsteke A, Verbeke W (2009) Public Good Markets: The Possible Role of Hybrid Governance Structures in Institutions for Sustainability. In: Beckmann V, Padmanabhan M (eds) Institutions and sustainability political economy of agriculture and the environment—essays in honour of Konrad Hagedorn. Springer Science and Business Media BV

Wade R (1987) The management of common property resources: collective action as an alternative to privatisation or state regulation. Cambridge J Econ 11(2):95–106

Wade R (1988) Village republics: economics conditions for collective action in South India. ICS Press, Oakland

World Bank (2006) Enhancing agricultural innovation: how to go beyond the strengthening of research systems. The World Bank, Washington

World Bank (2009) World Development Report 2010: development and climate change. The World Bank, Washington

Chapter 3
Collective Agri-environmental Strategies

Abstract The chapter provides a review of examples of collective agri-environmental initiatives that have been developed in Australia and in some European countries (Germany, the Netherlands), by showing their relevance in the current policy and academic debate. Although the cases described cannot always be classified as collective actions, an analysis of their main features provides some insights into the potentialities and barriers to implementing collaborative agri-environmental schemes in different contexts. Moreover, these examples reflect the complexity of the different agricultural and forestry practices as well as the difficult challenge of tailoring the collective strategies to local and regional needs. The final part of the chapter focuses on the main policy and institutional challenges for developing and supporting more grass roots collective actions for public goods, by providing a short overview of the Italian case studies that are analyzed in more depth in chapter 4 (the project Custody of the Territory in Tuscany Region) and in chap. 5 (the agri-environmental agreement in Marche Region).

Keywords Collaborative agri-environmental schemes · Cooperatives · Landcare · Grass roots collective action · Case studies

3.1 Introduction

The recent literature on the agri-environmental policies acknowledges the lack of effectiveness and efficacy of the approach currently in place. It is increasingly recognized that in many cases, for example in the context of the CAP, the current agri-environmental schemes (AES) usually do not encourage landscape level coordination, by favoring a farm scale approach leading to individual, disconnected actions (Prager et al. 2012:245). The lack of coordination in implementing agri-environmental strategies is strictly related to two additional limitations. Firstly, by preventing coordinated efforts and knowledge exchanges between farmers involved into the AES, in the majority of the cases such measures have failed to changes farmers' attitudes towards natural resource management (Burton et al. 2008; Burton and Paragahawewa 2011).

Secondly, the strategies targeted to individual farms bring out the problem of the 'scale mismatch' between the scale of administrative management of the measures

F. Vanni, *Agriculture and Public Goods*,
DOI 10.1007/978-94-007-7457-5_3,

(the farm scale) and the scale of the ecological processes which the schemes should manage (Cumming et al. 2006). Conversely, an improvement of water management, biodiversity and other ecosystem services through agriculture may be efficiently addressed only at landscape scale.

While conventional AES may be considered as top-down collective action cases, there is an increasing emphasis, both at EU and at extra-EU level that collaborative approaches are needed in order to increase the efficiency of such policies.

A recent study of the OECD (2012) on farmers' behavior reveals, for example, the importance of collective action in the context of mitigation and adaptation to climate change. Similarly, public goods such as biodiversity and landscape may be provided efficiently only by multiple persons and through collective action of farmers and other stakeholders in rural areas. At the same time, it is evident that collective agri-environmental strategies are more difficult to implement, because of several structural and organizational barriers related to transaction costs, as well as to monitoring and enforcement rules (see Chap. 2).

In spite of these limitations, it is increasingly recognized, also at institutional level, that in order to provide effectively agri-environmental public goods such as biodiversity and landscape, it is necessary to explore innovative solutions, with the joint involvement of farmers and of other rural stakeholders in the same area. Thus, an increasing interest on collective agri-environmental strategies can be observed, but in many cases local and national institutions fail to establish coherent incentives for the development of territorial agri-environmental measures. Indeed, cooperation is not so common in the environmental realm and it depends mainly on the local environmental context. The main challenge is establishing the conditions where groups of farmers within local areas can agree to adopt a coordinated approach to resources management. There is thus a need to identify appropriate areas and to set an agreed approach to management (Hodge 2001).

Nevertheless, in the international literature it is possible to find an increasing number of examples, including in developed countries, of collective approaches implemented for delivering agri-environmental public goods and/or for reducing the negative externalities.

For example, at EU level innovative schemes are emerging, whereby the use of a range of different measures (integrated delivery) is coordinated through landscape-scale delivery solutions, with the objective of increasing the efficiency of agri-environmental policies.

The next section includes a review of selected initiatives across the EU (Germany, the Netherlands) as well as in Australia. These examples were considered particularly relevant in order to provide a short overview of collaborative agri-environmental schemes (AES) that have been developed in different contexts.

Before describing these cases it is important to highlight that these collaborative AES are not always strongly related to the concept of collective action. Indeed, the collaborative AES are usually schemes that promote collective management and cooperation between farmers and between farmers and other actors, but not always are these examples the results of grass root collective action as defined in chap. 2. Nevertheless, these cases may provide some interesting insights on how to strengthen this collaborative approach and how to go further, by providing the financial, and institutional support for more bottom up collective actions that should scale up for increasing their environmental performance.

All the collective initiatives described in the literature (see Hodge and Reader 2007; OECD 2013; Poláková et al. 2011; ENRD 2012) show that there are different ways to deliver agri-environmental public goods collectively. At the same time, all these approaches usually interrelate and overlap and it is not easy to create divisions and to classify them. This variation reflects the complexity of the different agricultural and forestry practices as well as the need for tailoring the strategies to the local and regional needs.

Indeed, even though the examples described in the following sections shed light on the different approaches that, in different contexts, have tried to tackle agri-environmental problems at a territorial level rather than at individual farm level, the related policies were not always implemented with the specific target of promoting collective action in rural areas.

In sect. 3.2 three examples of collaborative agri-environmental schemes are described: the case of environmental cooperatives in the Netherlands, the Landcare Associations in Germany and the Australia Landcare Groups.

The last section of the chap. (3.3) is dedicated to less institutionalized and more grass roots agri-environmental actions, by introducing the Italian case studies that are analyzed more in depth in chap. 4 and 5.

3.2 Examples of Collabourative Agri-environmental Schemes

3.2.1 Environmental Cooperatives in the Netherlands

In the Netherlands a collective approach to agri-environmental policies has been traditionally ensured by the environmental co-operatives, which are formal organizational structures that offer to the members the opportunity to participate collectively in projects related to sustainable agriculture. These cooperatives, that may be defined as 'innovative associations of farmers based at local or regional level, which promote and organize activities related to sustainable agriculture and rural development in their locale' (Renting and Van der Ploeg 2001:87), also offer the opportunities to implement agri-environmental policies at landscape level.

These organizations involve mostly farmers, but usually also include various local and national agencies as well as environmental NGOs. This diverse range of stakeholders is a crucial aspect in adopting a pro-active approach on agri-environmental policies based on local needs. Indeed, the main goal of these local organizations is integrating nature management into farming practices by adopting a pro-active and inclusive approach based on a regional perspective (Franks and McGloin 2007).

The first environmental cooperative was established in 1992 as a self-help group with voluntary membership. During the decade 1990–2000 the number of cooperatives increased considerably, with the development of a broad range of collective initiatives related to several environmental issues, including biodiversity, landscape and heritage conservation. This increasing role has been traditionally supported by the Dutch government, which has been promoting a collective approach to AES since early 2000s.

Currently in the Netherlands there are around 150 regional groups, involving over 10,000 farmers and covering more than half of the Dutch countryside. The cooperatives' activities focus on conservation issues, such as birds of grassland and arable land, landscape features, but also on activities more related to rural tourism and farm education.

Environmental cooperatives are organized with an elected board of officers (e.g. chairman, treasurer, secretary, etc.) and follow their own rules (constitution), which set out specific procedures which govern the roles and functions of their members. Each cooperative carries out several projects, but there is a high flexibility that allows each member to participate in the projects that are interested in and this approach ensures only those projects with sufficient support go ahead (Franks and McGloin 2007).

As pointed out by Oostindie and Broekhuizen (2010), the emergence of these cooperatives may be also explained by the farmers dissatisfaction with available agri-environmental policies, since they have been claiming for more flexible, tailor made and financially attractive AES that may contribute to a larger extent to the professionalization of agriculture's provision of nature and landscape values. Indeed, in many cases these organizations function as intermediaries between public administrations and individual farmers, other landowners and rural dwellers interested in and committed to nature and landscape management, allowing Dutch conservation agencies to develop environmental management contracts with groups of land managers.

The Dutch government has given increasing support to the environmental cooperatives, by increasing their involvement in the definition and implementation of environmental policies at the national level, and by using them for experimenting and testing innovative policy tools.

Recently in the Netherlands AES have been implemented, thanks to the cooperatives, with the option of joint applications by neighbouring farmers with positive outcomes from both an environmental and social perspective. From an environmental point of view, the joint application in many cases allowed the overcoming of 'scale mismatch' highlighted above, since the adoption of sustainable farming practices at landscape level increases the connectivity between natural habitats, by ensuring a coherent management of natural resources at territorial level. The landscape level of the implementation of AES, ensured by the environmental cooperatives, is on a scale that is more appropriate for addressing ecological issues (Franks and McGloin 2006).

In terms of social outcomes, these innovative forms of organization helped to form trust-building local networks and, above all, allowed the integrating of farmers' views and knowledge into the design and the evaluation of agri-environment policies.

Indeed, even though such cooperatives generally accept and endorse policy objectives set by state agencies, usually they claim substantial reforms, more flexibility in implementation and advocate integrated development of land use and economic activities in their region (Van der Ploeg and Renting 2003). In many cases environmental co-operatives also represented interesting field laboratories for innovation towards sustainable rural development.

As highlighted by Franks and McGloin (2007), a fundamental requirement of co-operative actions is that individuals willingly work together and the national authorities are assisting environmental cooperatives to explore the most efficient institutional arrangements to overcome specific transaction costs (such as start-up costs) and to maintain clear incentives to collaborate (i.e., through the development of specific policy and regulatory tools).

These innovative arrangements and policy tools that may strengthen the role of environmental cooperatives have recently been the focus of attention, both at national and EU level, as possible solutions to increasing the effectiveness of the AES in the context of the 2014–2020 CAP programming period. More specifically, the Dutch Ministry of Economic Affairs, Agriculture and Innovation has selected four of these associations to experiment a regionally tailored approach towards ecosystem services and public goods provision. The associations involved are: (i) Water Land & Dijken Agrarian Nature Association; (ii) Agrarian Nature Association Oost-Groningen; (iii) Vereniging Noordelijke Friese Woudenand and (iv) Stichting Waardevol Cultuurlandschap Winterswijk.

These associations have developed ideas outlining a roadmap for the future CAP, not only from a rural development perspective, but also based on a holistic approach to land management for all regions and regional groups of farmers. Through these four CAP pilots, the ministry has investigated the possibilities of overcoming the practical problems involved in a collective approach to AES under the current public support (both form the CAP and form the national funding).

As already mentioned, CAP payments are currently directed almost exclusively at the individual farmers; in the Netherlands this approach has been traditionally perceived as an important bottleneck. Thus the objective of these four pilots' case studies was exploring the possibility that farmers, possibly in combination with other land managers, could be eligible for CAP payments as groups. Several official documents and papers of the National government highlight that a collective approach to AES could secure a more efficient and effective delivery of certain public goods and from a Dutch perspective the AES should be redesigned by establishing direct contracts with these cooperatives. This system should allow cooperatives to deal with the central government for the accountability, administrative and financial management of the contracts, and with the individual farmers regarding the management of individual contracts (payments, controls). According to the local experts (see Terwan 2012), this 'front door—back door principle' would increase the effectiveness of agri-environmental measures and, above all, it would simplify the system and reduce the administrative costs of the measures.

3.2.2 *Landcare Associations in Germany*

Landcare Associations (LCAs) in Germany are regional non-profit associations in which farmers, local administrations, politicians and environmentalists work together to implement nature conservation measures.

The first LCA was founded in Bavaria in 1986 with the objective of enhancing the collaboration between the opposing interest groups in rural areas, namely local farmers, nature conservation groups and politicians. The challenging tasks of LCAs was then bridging the gaps between different land use interests and coordinating the various stakeholders' actions related to the management of natural resources.

In more detail, the objectives of LCAs in Germany may be synthesized in the following ones (Blüemlein 2009):

- Create a network of natural habitats in order to protect the native fauna and flora as well as the biological resources in cultivated landscapes;
- Support environment friendly land use systems and regional economic development;
- Help farmers earn additional income from conservation and landscape management and market their regional products;
- Raise awareness of the importance of appropriate land management through the Landcare approach, involving the formal education system (even Kindergartens) to ensure that Landcare is mainstreamed in current and future education curriculum and programs.

Since the 1980s the activities of LCAs have strongly evolved. While initially such activities focused mainly on public lands and were related to planting trees and hedges to maintain the cultivated landscapes, more recently there has been an increasing involvement of farmers and LCAs promoting environmental friendly production systems in private lands, with the objective of promoting sustainable land use management but also to help farmers to commercialize their quality products.

Each LCA consists of members, a steering committee, a coordinator and a panel of external experts. Members of an LCA may be individuals, organizations, administrations, municipalities and private companies. The only requirement is that farmers, conservationists and local politicians should be equally represented on each steering committee. Usually the success of an LCA is closely linked to the presence of a coordinator who acts as a permanent contact person and the central actor for networking; not only with land managers and other concerned stakeholders, but also with administrations.

LCAs have also traditionally encouraged and supported individual farmers to adhere to the EU AES, by organizing and supervising the implementation of such schemes at the farm level. Landcare coordinators also monitor the schemes outcomes working closely with local communities, conservation groups, and government authorities such as conservation, agriculture, forestry and water management authorities.

LCAs rely on various financial funds, both from the federal state and municipalities and/or from CAP measures for the conservation of the rural heritage. Funds for carrying out land management are provided by membership fees and by EU-funded measures. In many cases LCAs operated as Local Action Groups (LAGs) for the EU LEADER programme. This may be considered a threat for further developing the approach of LCAs, since their future is somewhat dependent on the budget for LEADER and subsequent EU initiatives (Prager and Vanclay 2010).

To date, in Germany there are around 155 regional LCAs, involving more than 3,000 municipalities, 1,000 organizations and 20,000 farmers, and the majority of LCAs are concentrated in Bavaria (35 % of the total).

The national umbrella organization of LCAs is the German Association for Landcare (Deutscher Verband für Landschaftspflege, DVL) which provides technical and administrative support to single associations and supports the formation and development of new LCAs. DVL also communicates knowledge and experiences of different regional LCAs, improving the Landcare network at national level. Indeed, LCAs often collaborate closely with other LCAs to exchange expertise and experiences on good practice or to co-ordinate the management of a larger area. The DVL also works together with relevant ministries, represents the interests of LCAs at the European scale, and assists with building the foundations of new LCAs.

Even though the individual projects managed by the LCAs can be of different sizes, the added value of these projects is the co-ordination of measures at the landscape level; particularly important for biodiversity conservation and habitat connection. The majority of activities regard soil conservation, the management of hedges and meadows for habitat protection, the preservation of traditional orchards and the management of Special Protection Areas (Prager and Vanclay 2010).

From a social perspective, this collective approach experimented with LCAs has helped to develop trusting cooperative relationships with community and government and, above all, facilitated dialogue between conservationists and farmers, a crucial aspect to apply the principles of multifunctional agriculture at district level rather than at individual farmer level.

3.2.3 Landcare Groups in Australia

The European occupation of Australia in the 18th and 19th centuries brought a massive exploitation of natural resources, including biodiversity losses, salinity, weed and pest infestations and poor water quality.

During the period 1945–1985 the state and national governments have increasingly recognized the need for reducing the widespread land degradation of the country that in many cases was due to the introduction of the European style of agriculture and pastoralism.

In this period local and national institutions started to provide financial support to farmers for the conservation of natural resources, especially native ecosystems, but also for protecting soils, for improving the landscape and for reducing water pollution. In the frameworks of these programmes it was realized that there was a need for farmers to work together to have a more positive environmental impact at catchment level, and regional and integrated strategies were increasingly developed and supported. Thus, many territorial projects flourished with an increasing involvement of community organizations, and in 1986 a programme called 'Landcare' was launched by the government of Victoria State. This multi-disciplinary and community based programme, which aimed at capturing a broad spectrum of social and technical aspects in resource management, appeared particularly attractive to local communities. Thus, during the 1990s the approach rapidly spread in every state, also as result of the National Landcare Programme (NLP), which was launched in 1989 with the objective of developing effective partnerships and collective action between government authorities, local communities and industries (Johnson et al. 2009).

Since then, governmental policies for Landcare have included specific legislation based on sustainable development principles, and an Australian Landcare Council was established to provide grants and funding for local natural resources and management strategies. In the meantime a number of grant programmes have also been provided by local, state and territory governments with the same objectives.

Indeed, even though Landcare has based its financial support mainly from national and state governments, local government, environmental organizations and private consultants sometimes provide additional support or offer office spaces, administrative back up, meeting rooms, vehicles and funding.

Additional non-governmental support was also ensured by the National Farmers' Federation, by the Australian Conservation Foundation and by Landcare Australia Limited, the official not for-profit company responsible for seeking sponsorship for and promotion of Australian Landcare groups.

With regard to the professional support for Landcare, the government in 1990 also established a National Landcare Facilitator, to advise the government and to coordinate Landcare and other community groups. At the local level, similarly to the German Landcare Associations, many Australian Landcare groups recruited a paid coordinator to provide assistance, to organize meetings and to provide the management of the groups' activities.

The principle behind this complex set of national and local policies settled for Landcare groups are related to the increasing willingness of government to design programs that addressed the causes of natural resource degradation rather than their effects. The support to Landcare groups was considered the most effective solution to deal with the factors influencing farmers' management decisions, which include lack of knowledge or understanding, limited finances, attitude to risk or lack of skills (Willcocks 2008).

Indeed, the philosophy of Landcare groups is based on coordinated actions carried out on private land across a number of farms. This collective strategy involves government officials, NGOs and farmers and is finalized to provide a mix of private and public goods.

One of the key aspects of the Landcare approach experienced in Australia is widespread amongst the other collective and collaborative experiences observed in the EU member states: namely the possibility of incorporating social aspects into environmental actions. Indeed, through Landcare groups local communities have understood the benefits of joint action to analyze and solve local problems and, by providing social cohesion, the approach in many cases has contributed to the health and welfare of local communities (Australian Landcare Council 2012).

In Australia there are currently around 4,000 Landcare groups as well as a broad range of related community groups and networks using a similar approach (Australian Landcare Council 2012). Many Landcare groups are farmer-based but there are also urban, peri-urban and coastal groups.

Moreover, Landcare groups are increasingly taking part in broader networks managed by community boards to take a more regional approach to land and water issues and to coordinate agri-environmental policies at the most suitable scale. Thus, networks are now the major community link to all levels of government and industry for financial support and information (Johnson et al. 2009).

These networks have resulted in an increased capacity to deal with bureaucracy of groups as well as in an increased ability in discussing more complex concepts and in increasing the effectiveness of coordinating participants' time and resources allocations (Sobels et al. 2001).

At the same, as recently highlighted by Prager and Vanclay (2010), despite the rhetoric of 'bottom-up', this structural coordination by the central government resulted in a remarkably top-down approach and this is one of the most relevant differences with the LCAs in Germany, where groups and networks are usually formed without government involvement.

In spite of this limitation, the approach developed in Australia may be considered a relevant and innovative model of implementing integrated and collaborative agri-environmental policies, based on the involvement of the different government tiers and a of broad range of private and public stakeholders. From a collective action perspective, the most relevant implication of the Australian Landcare initiative is related to the awareness of the different governmental tiers that only a broad partnership between public institutions, communities, farmers and education bodies can overcome the problem of land degradation and improve the sustainability of agricultural practices.

3.3 Grass Roots Collective Actions

As observed in the initiatives described above, agri-environmental policies in some cases may support collaborative projects, for example by providing funding that may be used for collective activities or, more likely, to promote more complex programmes where a broader range of stakeholders are involved. At the same time, it must be recognized that these collaborative agri-environmental schemes do not necessarily promote collective actions, since in many cases not all the stakeholders involved have the motivations to act collectively but are individuals that applied for collective schemes, which usually are settled with a rather top down approach.

Conversely, in many contexts, the need is emerging, in order to increase the effectiveness of agri-environmental policies, for also taking into consideration the role of more local, informal and grass roots collective action for public goods.

The existing examples in the literature demonstrate that, even though usually at a smaller scale compared to the collaborative AES, informal actions may have the potential for scaling up, by engaging a number of different actors a in a larger area or the coordination of a range of different stakeholder groups. In these cases usually local organizations act as intermediaries for actors between the local and the national level, in order to support planning and implementation of local projects with an environmental focus. In many cases civil society associations are the typical initiators of innovative development steps and their involvement usually contributes to a great extent to the success of such initiatives.

More specifically, the existing examples at the EU level based on community-led and collective approaches demonstrate that such grass roots action (ENRD 2012):

- Can help to provide flexible and local strategies to deliver environmental public and services;
- Require coordination through some form of administrative bodies, for example nature conservation agencies, or local authorities;
- May have a wide range of additional benefits, including providing added value to agricultural products, promoting local tourism and increase awareness of environmental issues;
- Usually require significant investment in communication and advice activities to ensure engagement by the different actors and this can also lead to increased administrative burdens.

Thus, while the importance of making use of the high potential of such grass roots collective actions is increasingly acknowledged, it is evident that there are also important limitations, mainly related to the barriers of collective action, namely free riding and transaction costs (see sect. 2.2).

In order to address these barriers the availability of adequate financial resources is usually a key factor, since the free riding problem may be addressed through investment on monitoring and enforcement, while transaction costs are related to the necessary advice, information and communication. Nevertheless, less institutionalized agri-environmental actions in many cases may rely on informal monitoring and enforcement by group members that in some cases would also be necessary in both state and market solutions (Petrick and Gramzow 2012: 2343). From this point of view, as it will be observed in the case studies from Italy analyzed in this book, in grass roots collective actions traditional incentive mechanisms such as trust, solidarity, reciprocity and reputation are typically employed to reduce free riding.

The following chapters provide a more detailed analysis of these initiatives that relates to two different types of collective actions promoted respectively to increase the provision of environmental services and to decrease the negative externalities from agriculture (see Table 3.1).

The first collective action analyzed (chap. 4) is related to a local project in a mountain area of Lucca and Pistoia provinces, in the Tuscany region named 'Custody

Table 3.1 The Italian case studies

Case study	Custody of the territory	Valdaso agri-environmental agreement
Type of collective action	Collective action led by the local agency in charge of the project (coordination)	Farmers-led action, later institutionalized and supported by local institutions (cooperation)
Objectives	Increasing farmers' stewardship, landscape management, hydro-geological management, reducing farm abandonment	Reducing the environmental externalities of agriculture through the adoption of more sustainable practices (Integrated Pest Management)
Location	Media Valle del Serchio (Pistoia and Lucca Provinces, Tuscany) Mountain and marginal area	Aso Valley (Ascoli and Fermo provinces, Marche region) Area characterized by intensive agriculture (fruit production)
Public goods	Hydro-geological management of the territory	Soil quality, water quality and food safety
Policy measures	Local initiative, funded by a local agency and by the regional RDP	Measure 111 and 214 of Rural Development Programme
Approach to delivery	Holistic approach: local strategy for sustainability outcomes	Integrated delivery: package of RDP measures

of the Territory'. The project was created and implemented by a local government agency, which set an agreement with the local farmers for the co-production of the environmental services, in order to increase the resilience to flooding and to improve the landscape and hydro-geological management of the territory.

The second case study (chap. 5) focuses on the Aso Valley area (Valdaso) in the Marche Region, where a group of farmers started a grass roots initiative to adopt integrated management techniques at territorial scale, with the objective to protect water and soils from pesticides and nitrates pollution. This action, co-managed by the local farmers and by the local institutions, was supported through an 'agri-environmental agreement' (AEA) which ensured the delivery of public goods at landscape level.

As described in the Appendix, the collection and analysis of data for the two case studies were carried out through Rapid Rural Appraisal (RRA) methods. The participatory methods of the RRA were considered particularly suitable to establishing direct rapport with the respondents, with the objective of having the insiders' views and the understanding of the local actors involved in the collective actions under study. In particular, semi-structured interviews have been the main tool used to understand the relevant institutions involved as well as the rules, norms and attitudes involved in the collective actions. Moreover, other RRA techniques such as sketch maps and transect walks were also used, especially for identifying and analyzing more in depth the practical problems of the collective actions on the field, with a deeper learning which allowed reducing the gap between theory and practice.

The two case studies are quite different, since different territories and farming systems are involved, but also the type of grass roots collective action is quite differentiated in terms of institutional organization, approach to delivery and in terms of environmental objectives to be achieved. These differences allowed for the exploration of the role of collective action in different contexts and for different goals, in one case for providing environmental services such as the hydro-geological management and the resilience to flooding in mountain areas, and in the second case in controlling negative externalities in intensive farming systems.

Finally, as will be further discussed, these initiatives provided some useful indications on the policy tools as well as on the institutional arrangements necessary to promote local collective action for provision of the agri-environmental public goods.

References

Australian Landcare Council (2012) Australian framework for landcare. Department of Agriculture, Fisheries and Forestry, Canberra

Blümlein B (2009) Landcare in Germany. In: Catacutan D, Neely C, Johnson M, Poussard H, Youl R (eds) Landcare local action global progress. World agroforestry centre, Nairobi.

Burton RJF, Kuczera C, Schwarz G (2008) Exploring farmer's cultural resistance to voluntary agri-environmental schemes. Sociol Ruralis 48(1):16–37

Burton RJF, Paragahawewa UH (2011) Creating culturally sustainable agri-environmental schemes. J Rural Stud 27(1):95–104

Cumming GS, Cumming DHM, Redman CL (2006) Scale mismatches in social-ecological systems: causes, consequences and solutions. Ecol Soc 11(1):14. http://www.ecologyandsociety.org/vol11/iss1/art14/. Accessed 2006

ENRD (2012) Coordination committee focus group delivery of environmental services, final report December 2012, Brussels

Franks JR, McGloin A (2006) Co-operative management of the agricultural environment. SAFRD, Newcastle upon Tyne

Franks JR, McGloin A (2007) Environmental co-operatives as instruments for delivering across-farm environmental and rural policy objectives: Lessons for the UK. J Rural Stud 23(4):472–489

Hodge I (2001) Beyond agri-environmental policy: towards an alternative model of rural environmental governance. Land use policy 18(2):99–111

Hodge I, Reader M (2007) Maximising the provision of public goods from future agri-environment schemes. Final report for Scottish natural heritage, rural business unit, Department of Land Economy, University of Cambridge

Johnson M, Poussard H, Youl R (2009) Landcare in Australia. In: Catacutan D, Neely C, Johnson M, Poussard H, Youl R (eds) Landcare: local action global progress. World agroforestry centre, Nairobi

OECD (2012) Farmer behaviour, agricultural management and climate change. OECD publishing, Paris

OECD (2013) Providing Agri-environmental Public Goods through Collective Action. OECD Publishing, Paris http://dx.doi.org/10.1787/9789264197213-en

Oostindie HA, Broekhuizen RE (2010) RDP and agriculture's provision of green and blue services (GBS) in the Netherlands. Rudi project, Wageningen University and Research Centre, Chair of rural sociology, Netherlands

Petrick M, Gramzow A (2012) Harnessing communities, markets and the state for public goods provision: evidence from post-socialist rural Poland. World Dev 40(11):2342–2354

Ploeg JD van der, Renting H (2003) Environmental co-operatives reconnect farming, ecology and society. In: Haverkort B, Hooft K van 't, Hiemstra W. Ancient Roots, New shoots: Endogenous development in practice

Poláková J, Tucker G, Hart K, Dwyer J, Rayment M (2011) Addressing biodiversity and habitat preservation through measures applied under the common agricultural policy. Report prepared

for DG agriculture and rural development, Contract No. 30-CE-0388497/00-44. Institute for European Environmental Policy, London
Prager K, Reed M, Scott A (2012) Encouraging collaboration for the provision of ecosystem services at a landscape scale Rethinking agri-environmental payments. Land use policy 29(1):244–249
Prager K, Vanklay F (2010) Landcare in Australia and Germany: comparing structures and policies for community engagement in natural resource management. Ecol manage restor 11(3):187–193
Renting H, Ploeg JD van der (2001) Reconnecting nature, farming and society: environmental co-operatives in the Netherlands as institutional arrangements for creating coherence. J Env Policy Plan 3(2):85–101
Sobels J, Curtis A, Lockie S (2001) The role of landcare group networks in rural Australia: exploring the contribution of social capital. J Rural Stud 17(3):265–276
Terwan P (2012) Collective agri-environmental contracts: the Dutch experience. European network for rural development, 9th meeting of the coordination committee, Brussels 14 June 2012
Willcocks C (2008) The power of partnerships. In: Making a difference, a celebration of landcare, Department of Agriculture, Fisheries and Forestry, Canberra

Chapter 4
The project "Custody of the Territory" in Media Valle del Serchio (Tuscany)

Abstract This chapter describes the collective action related to the 'Custody of the Territory' project, an initiative promoted by a local authority of a mountain area of Tuscany ('Media Valle del Serchio') which set an agreement with local farmers for the co-production of some environmental services. This local authority set a Payment for Environmental Services (PES) with local farmers, with the main objective of increasing the resilience to flooding by improving the landscape and hydro-ecological management of the territory. This innovative approach to PES, based on collective action, shows how the relations between farming activities and environmental services could be addressed at territorial level, which resources are mobilized, what type of information is exchanged and what outcomes are reached. Finally this case study shows how social learning and co-production of knowledge (amongst farmers, institutions, technicians, and citizens) are very important issues for a collective provision of environmental services.

Keywords Payments for Environmental Services · Hydro-geological management · Mountainous areas · Co-production · Joint learning

4.1 Case Study Overview

During the last decade, an increasing frequency of flood events in large parts of Europe has been observed, partly as result of climate change. Nevertheless, the estimates of changes in flood frequency and magnitude remain uncertain (EEA 2008). Extreme weather events impact to a varied extent on different rural areas across Europe, but it may be argued that risks are exacerbated in those areas where, also as a result of agricultural intensification, rivers and other watercourses have been straightened, which increases the speed of the water's flow, where riparian vegetation has been removed, and on highly compacted soils leading to fast rainfall run-off (LUC 2009). Agriculture, however, can also form part of a mitigation strategy, with certain forms of land management improving the water storage capacity of the land and hence, the resilience of the broader landscape to these risks (Cooper et al. 2009).

In Italy, with regard to the magnitude of the hydro-geological risks, a study released by the national council of geologists (CNG 2010) highlights that 6 million inhabitants (about 10 % of the national population) live in the 29.500 km squares that

F. Vanni, *Agriculture and Public Goods*,
DOI 10.1007/978-94-007-7457-5_4,

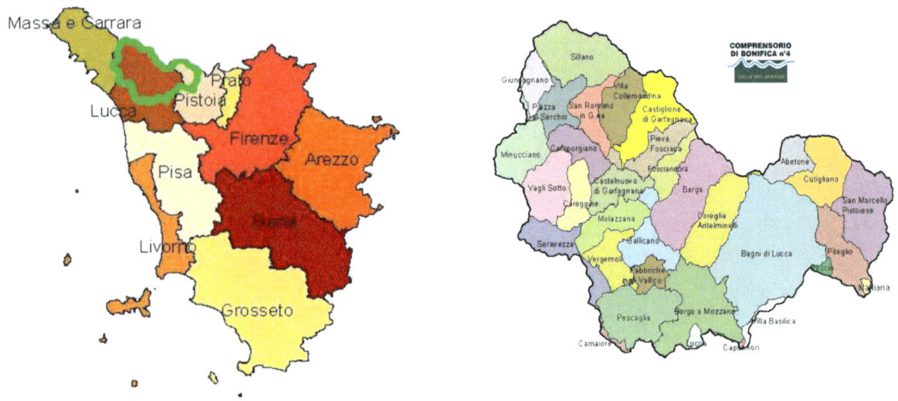

Fig. 4.1 The Media Valle del Serchio district, in Tuscany. (Source: Comprensorio di Bonifica Media Valle del Serchio)

are considered at 'high risk', while 1–2 million of buildings are considered at 'high risk' of landslides and flooding.

Tuscany is one of the Italian regions that during the last decade have been increasingly subjected to extreme weather events and flooding. According to the data released by Legambiente in 2011, 98 % of Tuscan municipalities are affected by phenomena of hydro-geological risks, which comprise the 90 % of the regional houses and infrastructures.

Thus, the hydro-geological management of the territory is increasingly recognized as a regional priority, especially in relation to the improvement of the resilience to flooding, and the role of farmers and other rural stakeholders may improve flood prevention management.

This is particularly true in mountain areas, where the management that farmers can undertake to reduce the risk of flooding on their own land may also reduce the risk of flooding at landscape level. At the same time, in many mountain areas of the country an increasing abandonment of farming activities has been observed, which resulted in increasing difficulties for the local institutions to ensure an adequate hydro-geological management of the territory.

The case study analyzed here refers to a local project in the District No. 4 'Serchio Valley', a mountain area of over 115,000 ha in the drainage basin of the Serchio River, in Lucca and Pistoia provinces of Tuscany region, which comprises 35 municipalities (Fig. 4.1).

From a socio-economic point of view, the area is characterized by low population density, mainly as result of a significant process of abandonment of the most marginal and remote areas, especially in the mountains. This phenomenon resulted in a higher urbanization process in the plains, which are characterized by high population density and high presence of infrastructures. It is then possible to observe a strong dualism between the areas of the plains, where the industrial and residential activities are concentrated, and the mountain areas, where agricultural and forestry activities still play an important role (Fig. 4.2).

Fig. 4.2 The dualism between the plain and the mountainous areas of the district. (Source: Comprensorio di Bonifica Media Valle del Serchio)

For these reasons, in this district the hydro-geological management of the territory is increasingly recognized as one of the main environmental priorities, also as a result of the numerous extreme weather events that, during recent years have increased the perception, amongst the local citizens and politicians, that the area needs a stronger resilience to flooding.

The high risks of over-flooding are related to the strong development of residential, industrial and manufacturing activities in the plains as well as to the high hydro-geological fragility of the local mountain areas, due to the characteristics of the soils and the high level of yearly rainfall, but also to the lack of landscape and environmental maintenance activities.

In this district, the main reclamation activities are managed by a local authority, the Mountain Community 'Media Valle del Serchio', which is in charge of ensuring the hydro-geological management of the territory (the agency is in charge of cleaning up and restoring the riverbeds, as well as maintaining the 2,500 hydraulic structures of the areas, mainly dikes).

This local government authority, due to difficulties in managing over 115.000 ha of mountain areas and about 1.500 km of streams and torrents, has decided to set a Payment for Environmental Services (PES) with local farmers, with the main objective of increasing the resilience to flooding by improving the landscape and hydro-geological management of the territory.

PES were defined by Engel and Palmer (2008) as 'voluntary transactions where a well-defined service is bought by a buyer from a service provider if and only if the provider secures its provision (conditionality)' (Fig. 4.3).

The PES for the local farmers settled by the local authority in Media Valle del Serchio is articulated into two types of activities:

- Monitoring activities: periodical on site controls of torrents and streams, with reports and pictures;
- First maintenance intervention: execution of simple maintenance works such as removal of trees, wood and debris from riverbeds and dikes to avoid overflowing, together with the management of riparian vegetation (Fig. 4.4).

The idea beyond this strategy is that these type of activities, which have been increasingly dependent on the activities of specialized firms, could be carried out more efficiently by local farmers living and working in the district.

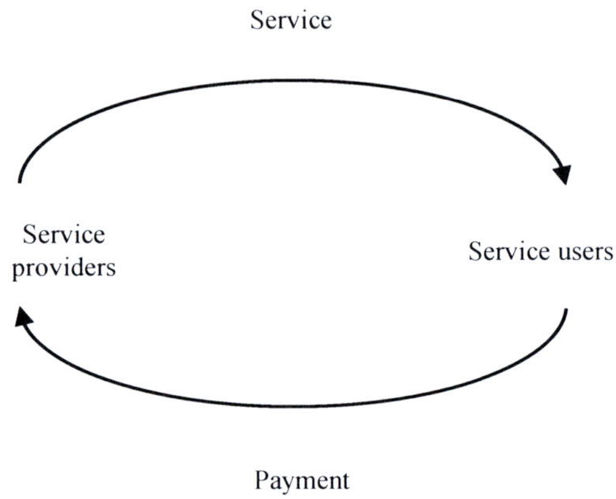

Fig. 4.3 Payments for environmental services

The assignment of these types of environmental services by the local authority is usually the result of the monitoring carried out by farmers, who prepare a detailed report on the canals requiring the intervention works, with their location, as well as with a short explanation of the needed works and some pictures. If the service to be provided can be carried out by the farmers (cleaning up riverbeds and other simple operations) who have signalled it, the local agency assigns directly the environmental services to the farmers.

The PES includes a fixed payment (€ 6.000 per year during the initial phase and € 4.000 per year during the following years) for the monitoring activities, and a variable payment for the first maintenance intervention, based on the extent of the work to be done. The maintenance activities, according to the Italian law on multifunctional agriculture and diversification activities (national Legislative Decree n. 228/2001), cannot exceed € 50.000 per year for professional farmers and € 300.000 for specialized cooperatives.

During the initial phase of the project (years 2007–2008), the initiative has received great interest from farmers: there were 63 expressions of interest that resulted in the activation of 20 contracts with 'farmer custodians'. In this phase the activities

Fig. 4.4 Monitoring activities and first maintenance intervention

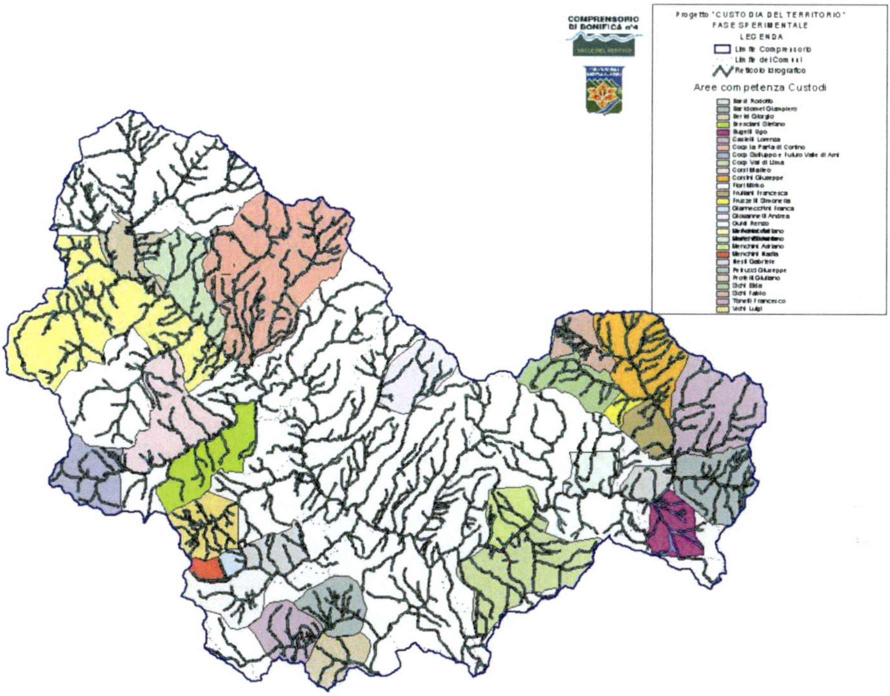

Fig. 4.5 The area 'covered' by 'farmer custodians' in 2011. (Source: Comprensorio di Bonifica Media Valle del Serchio)

were mainly related to the monitoring (report writing and pictures), since it was necessary to develop a data-base of the main environmental needs of the area and of the environmental services to be carried out by farmers.

During the second phase (years 2009–2010), some organizational aspects of the project were modified, and some 'farmer custodians' were replaced for the exigencies of the local agency in terms of territory to be covered by the agreement (and in some cases for the lack of farmers' professionalism), and 13 new farms were involved in the agreement. In this phase the budget was decreased for the monitoring activities that, logically, had been carried out extensively during the first phase.

During the third phase (2010/2011) the PES were settled with 25 farmers and 4 cooperatives. In 2011 the local agency was then able to monitor 500 km of torrents and streams, corresponding to the 40 % of the territory (Fig. 4.5).

The broad objectives of the project may be summarized as follows:

- Improving the environmental management of the areas through the involvement and empowerment of local communities;
- Favoring a pro-active role of farmers in managing the territory in order to maximize their role in delivering the environmental services;
- Increasing the resilience to flooding by favoring the involvement of farmers in prevention activities (monitoring, surveillance, early intervention works).

4.2 Local Farmers as 'Custodians of the Territory'

4.2.1 A Knowledge-based Scheme

The recent literature on policy science emphasizes how the technocratic and centralized knowledge making is failing to solve the most pressing social and environmental problems. For this reason, it is generally argued that public administration should rely more heavily on local and lay knowledge (Corburn 2007).

From this perspective, one of the most important innovations of the 'Custodians of the Territory' project is related to the identification of farmers to be involved in the initiative, since they were selected not only on the basis of economic and technical parameters, but also on the basis of the location of their farms and on the basis of their knowledge of the territory.

Indeed, the first phase of the project was a mapping activity, carried out by the local authority in cooperation with the municipalities of the district, aimed at identifying the areas 'covered' by the presence of farmers. This mapping activity was supplemented by a direct survey among local farmers through interviews, with the objective of understanding their knowledge of the territory and, above all, their 'range of competence'. As put by one of the initiators of the project:

> We did not investigate only farmers' knowledge, but also their "range of knowledge". We asked local farmers the following questions: even though your farm is only 15 ha of forest and arable land, in reality how do you move around? Where do you go usually? How far do you feel that you the knowledge of the territory? How often do you go in these areas? They explained to us their displacements and above all their knowledge of the areas surrounding their farms and this was crucial to settle the agreement for the environmental services that they could provide (INTERVIEW n. 1).

After the identification of possible farmers to be involved, the local agency gave recognition to and a mapping of the environmental priorities for the hydro-geological maintenance of the territory and then drew the first agreement, in consultation with the local farmers' organization (Coldiretti and CIA).

This sound, context specific socio-ecological research prior to the implementation of the project is based on the assumption that the design of a PES schemes should be context-specific, due to the uneven information of socio-ecological systems (Muradian et al. 2010).

According to respondents, the real added value of the environmental services which are provided by the farmers is linked to their local knowledge of places (location and conditions of canals, streams and hydraulic structures) but also in a more broad knowledge of the territory in terms of local people, traditions and history.

According to the stakeholders interviewed, the type of local knowledge that plays a crucial role in the project is related to the folk memory of farmers regarding the environmental dynamics of the territory. This folk memory is usually passed down orally over the years and it involves, for example, the knowledge about the extraordinary meteorological events of the past, the seasonal (and annual) water level of rivers and streams, the diversion of water flows, the flow of water in times of flood, the access points to streams and rivers.

The local knowledge of farmers related to the spatial and temporal dimensions of environmental priorities (i.e. the risk of flooding) was considered crucial to identifying the main environmental risks, as well as the timing and the location of the interventions.

From this perspective, the project was encouraging preventive actions, especially in the most remote and inaccessible areas of the district. This was possible through the daily monitoring and local knowledge of local farmers. Indeed, the economic incentives to the monitoring allowed for the creation of a knowledge network of farmers that keeps the areas surrounding their farms under control, an area of expertise based on their location and on their commitment to resolve the emerging environmental problems of these areas. Indeed, one of the most innovative aspects of this project involved local farmers carrying out the monitoring and maintenance works not only within the farms' boundaries, but also on an area surrounding their farms. This widened the spatial perspective of many farmers, by increasing their perceptions of the territory in which they operated in terms of landscape and hydrogeological problems. This aspect was well synthesized by a farmer involved in the project:

> I believe that the fact that the project pushes us to go out from our farms is very important, because when we deal with canals and streams we cannot think about farms boundaries. The hydro-geological management of the territory is not a business problem, is a common good and this project allows us to better understand that maintaining common goods is an important opportunity. The main goal is increasing the participation of local farmers, who should also demonstrate to be keen and active for this activity. With regard to my feeling, I have to say that I feel this as my territory, where I was born and I have my roots. I carry out the monitoring activities and the intervention works also for these reasons, for my territory and not just for my farm (INTERVIEW n. 11).

The monitoring activity can be considered the most innovative aspect of the project, and it is intrinsically linked to the farmers' deep knowledge of the area. In fact, according to the stakeholders interviewed, only people living in the territory are able to provide fruitful indications about the exact location of the environmental problems and of the flooding risks that occur in such areas. The initial phase of the project was then based on recognition of this knowledge amongst farmers, which has been considered crucial to identifying the types of issues to be solved and their locations, in order to increase their efficiency and effectiveness of the collective action.

Moreover, the data collected show that the local knowledge of farmers was also very useful to deal with the administrative aspects. For example, the direct knowledge of landowners by farmers was a crucial step in planning interventions, since it has been shown to play an important role in facilitating administrative procedures, by reducing the red tape and by avoiding possible contestations or other legal problems.

On the basis of this evidence, it may be argued that, while in many cases professional and government representatives fail to recognize the different contribution that local knowledge can make to problem solving, through the 'Custody of the territory' project the local decision makers aimed at re-valuing and re-constructing different forms of local knowledge which were previously excluded from the main strategies regarding the management of the territory. These forms of knowledge were key factors in helping local farmers to identify risks and in setting the early intervention on the canals and streams in order to prevent flooding.

Finally, this approach promoted the spread of best practices among farmers, recovering the daily actions of prevention and maintenance that in many cases had been lost. Interviewees agree that, since the territory is no more largely occupied by farms, an inclusive collective action is the only solution to ensure an effective delivering of environmental services at landscape level. As pointed out by a member of a local municipality:

> The abandonment of farming determined many environmental problems, especially in relation to the landscape maintenance works, which in the past were naturally carried out that all the farmers that were settled here, who by maintaining their land in good conditions also maintained the all territory. Nowadays the problem is different; the main challenge for local authorities is encouraging a collective action of farmers, who must provide services outside their farms. This means to have effects on farmers' motivations and attitudes, in order to recover their knowledge and their competencies in providing environmental services which are not only functional to their business (INTERVIEW n. 3).

From this perspective, as discussed in the next section, the interviews with local farmers have allowed for the exploration of how they perceive the environmental services that they carry out, how such services fit in their farming strategies and which are the main drivers that push them to adhere to the project.

4.2.2 Motivations and Attitudes: Building the New Identity of 'Farmers Custodians'

The data collected in the context of the present study confirm what several economists have already demonstrated: in many cases the participation into collective action is strongly motivated by private benefits. Indeed, farmers adhered to this project mainly because they believed that they would gain from participation both directly, by having access to the specific funding for monitoring and for carrying out the first maintenance works and indirectly, by increasing the opportunities to also collabourate with other government authorities.

In addition, according to the coordinator of the project, since the possibility of first intervention works is led by the monitoring activities, in some cases farmers tend to exceed in signalling the need for intervention in order to increase the possibility to carry out the first intervention works.

In fact, in some cases, farmers may decrease the monitoring activities when the local authority does not assign the first intervention works to them, either because of lack of available funding, or because the work is of a higher entity (in terms of machinery available and competencies) and is to be assigned to specialized companies or cooperatives.

At the same time farmers have also emphasised that they perceive this type of PES as an efficient way to optimize and rationalize farms activities (use of labor, machineries and spare time) but also an interesting opportunity to increased their visibility and reputation.

Local farmers usually carry out the monitoring activity that during the idle time, since it represents an interesting complement to productive activities. Also because it allows engaging employees at a time when there is a lack of activity in the farms. In many cases this activity fits well with the hobby activities of farmers such as hunting, fishing and mushroom picking. Similarly, the first intervention works were considered important income integration in periods of scarce activities in the farms (i.e. during the winter and in raining days).

Nevertheless, as observed by Muradian et al. (2010), the economic incentives are just one of the main drivers that may influence farmers' behavior and farmers' willingness to adhere to the PES. The data collected on the field confirmed this, showing that the motivations and attitudes of farmers usually go well beyond the private and economic interests, and are also related to the personal sphere of farmers and to their identities.

For many farmers the reasons for joining the project are related to their personal passions, skills and ideas and, above all, as they emphasized during the interviews, to the fact that the project contributed to reinforcing their identity as farmers.

For example, many of them expressed their concerns about the increasing farming abandonment of those marginal and mountain areas, and they perceived the project as an opportunity to increase the recognition for their role of landscape stewards. As put by a local farmer:

> There is also some pride in participating in this project. We can change something in people conscience and awareness. When people see a farmers working outside his farms for preventing hydro-geological disasters and flooding they may think that there is something important to think about, namely the environment, and maybe farmers may help us to increase the safety and the beauty of our territory (INTERVIEW n. 13).

Some farmers emphasized that their participation into the project makes them feel directly involved in the management of the territory, and this is view also as a social role, since the services provided may increase local people awareness on the importance of the farmers' stewardship in mountain areas.

Another interesting issue that emerged during the interviews was the qualitative aspect of the activities carried out by the local farmers compared to other local actors who potentially could be involved in the PES. Indeed, the approach of farmers in delivering the environmental services was described by the representative of the local authority as 'different in qualitative terms' compared to the approach of the authority's workers or compared to those working for specialized cooperatives, since usually farmers in delivering the environmental services put the same commitment and dedication into their farming activities.

Indeed, while the approach to the work of public or private employees are mainly regulated by the contract of employment (with annual leave, work permits and other types of permits), farmers who have joined the project consider the environmental services as one of their daily work activities within their farms. For example, respondents noted that, compared to the other types of workers, farmers are used to complete the works as soon as possible, also in bad weather conditions and after working hours.

Box 4.1 Transect walks with the 'farmer custodians'

Transect walks represented a good integration and continuation of the semi-structured interviews carried out with the 'farmer custodians'[1]. During the transect walks farmers explained the main hydro-geological problems of the district, with detailed descriptions on the conditions of canals and rivers observed alongside the walks.

Farmers simulated these activities and this gradually led to the identification of the main problems that otherwise would not have been addressed. These transect walks also provided an opportunity for discussion on the main technical problems encountered by farmers during the monitoring activities and during the intervention works.

Farmers also described in detail the environmental and topographical features of the area in which they operated, the problems of access to the main canals, the machineries and the knowledge needed to carry out the intervention works, the services that have been already provided and those that are planned to be provided. The data collected was related to both the technical knowledge required to carry out the environmental services (i.e. how positioning the wood alongside the rivers and the canals after the cutting operations) but also to the more general views (i.e. motivation and attitudes towards the project).

[1] The transect walk ended with summarizing the information gathered during the walk and this was used as a point of discussion for the following interviews and transect walks, favouring the 'iteration' process of the grounded theory approach undertaken in the research (see Appendix).

Table 4.1 Farmers' motivation to participate into the collective action

Individual/economic reasons	Social/environmental reasons
Payments for monitoring and first intervention works	Enthusiasm and keenness for the environmental services to be provided
Interesting complement to productive activities	Contributing to the environmental management of their territory
Engagements of employees when there is lack of activities in the farms	Social and institutional recognition of farmers as landscape stewards
Integration in periods of scarce activities in the farms (i.e. during the winter and in raining days)	Building a new identity of 'farmers custodians'
Increasing visibility and networking (opportunities to collabourate with other government agencies)	Involvement of farmers in decision-making process

This different approach in delivering environmental services by farmers also resulted in a different length and quality of the works, as emphasised by the municipal authorities interviewed who have benefited by the PES in their municipalities.

Furthermore, the data collected suggests that this approach, especially in the case of the most active farmers in the project, favored the building of a new identity of the 'farmer custodians'. This new identity was created as result of the social and institutional recognition of their strategic role as stewards of the local landscape and of the hydro-geological management of the territory.

Indeed, according to technicians of the local authority, the recognition of farmers as 'custodians of the territory' was interpreted by them as a sort of 'right' to make the monitoring and maintenance activities in their assigned area. This institutional recognition led local farmers to express the local needs in terms of flood prevention and landscape management activities and, more broadly, to be an example for the other local land managers and land owners.

This new identity was built, on one side, through the institutional recognition of their role and, on the other side, through a social acknowledgement, since local communities were increasingly aware of the role of local farmers in the management of the territory.

This recognition also resulted in an increasing involvement of farmers in the project decision-making, due to the new relationships, based on trust and cooperation developed amongst the farmers, the technicians and the responsible of the local authority.

This role and this recognition, in some cases, pushed farmers to increase the range of environmental services provided, such as removal of illegal dumps in riverbeds and in adjacent areas, even though these services are not included in the agreement with the local authority (and so without remuneration).

Finally, in many cases the most pro-active farmers became a point of reference for local people and, above all, represented a reliable information network as well as an efficient early warning system for the public administrations in charge of the environmental management of the district. All these materials, immaterial public and common goods are strongly linked to the collective action undertaken and to the social and cultural dimensions that have been mobilized through the project (Table 4.1).

4.2.3 Co-production of Knowledge and Learning Opportunities

As discussed above, the PES experience in the Serchio Valley area may be considered a collective action led by a local authority that settled a specific agreement with farmers to achieve joint goals, such as reducing the costs for the landscape maintenance works, but also to increase their effectiveness. These objectives were pursued through 're-building' the technical knowledge of farmers related to the environmental management activities. Regarding the technical aspects concerning the monitoring and intervention methods, respondents argued that the value added to the project is the exchange of knowledge and the learning experiences that have been activated.

Indeed, the project aimed at recovering this knowledge through the interactions and exchanges between different actors (institutions, technicians and farmers), in order to increase the effectiveness of the environmental services provided. Instead of implementing the traditional hierarchical approach of learning transmission (manager to technicians to workers), the involvement of farmers led to a constructive exchange and to a learning process that has involved representatives of local institutions, technicians and local farmers.

More specifically, the interactions amongst farmers and the local agency technicians resulted in a process of learning and co-production of knowledge, which represented, on one side, an efficient early warning system for the risk of flooding and on the other side, ensured the provision of cost-effective environmental services.

These interactions are ensured, as a first step, by joint inspections carried out by the technicians working for the local authority and the farmers before the assignment of the environmental services, which is considered a crucial moment of knowledge exchange between farmers and technicians.

According to the farmers and technicians interviewed, the modalities regarding the provision of environmental services are the result of a useful integration of local knowledge and technical knowledge, since the evaluation of the environmental services to be provided, the type of works necessary and the time and budget issues are the result of a negotiation process amongst the technicians and the farmers.

On one side, technicians' knowledge and experience were mainly used for the intervention works, since this knowledge is necessary to evaluate the costs, the time and machineries necessary to carry out the services. On the other side, the farmers' knowledge was particularly useful for the identification of the environmental priorities, for the localization of the points of the canals and streams to be controlled, and, above all, to have the perception of the environmental risks. Thus, local stakeholders argued that the integration of these two types of knowledge have created a virtuous circle, a sort of osmosis of knowledge that enabled both technicians and farmers to learn about new concepts and issues.

According to the farmers, the environmental services required by the project are not particularly difficult to carry out, since they are used to make similar maintenance works for the canals and streams that are within the boundaries of their farms. At the same time, the project coordinator argued that in many cases the approaches of farmers were not appropriate, since they tended towards cutting too much of the

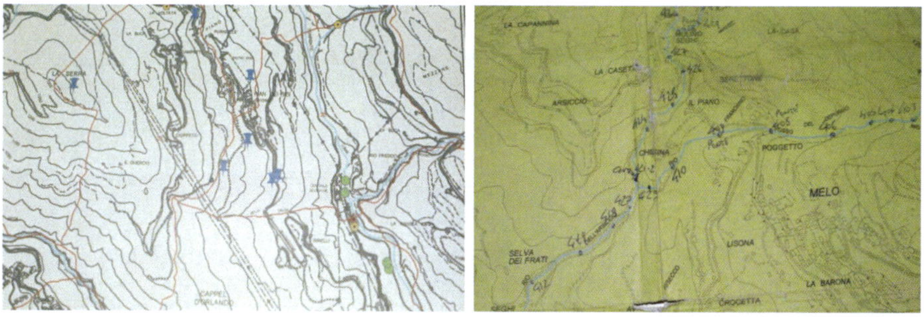

Fig. 4.6 The census of canals and hydraulic structures

riparian vegetation. Some farmers do not fully understand that many environmental services are complex operations, and are not simply the removal of vegetation and trees but are related to the management of complex and fragile ecosystems.

From this point of view, the project coordinator argued that it would be appropriate to provide specific training to farmers in order to carry out more efficiently all the environmental services required. Despite these critical aspects, the coordinator also acknowledged that in many cases the environmental services were carried out efficiently and at little cost, by cleaning entire sectors of streams and canals that were badly maintained for many years.

The interactions between farmers and technicians, both during the joint inspections and during the execution of the first intervention works are very important moments of exchange, in order to overcome the main technical difficulties encountered by farmers. According to the local stakeholders, the integration of local knowledge with technical knowledge was also the result of an inclusion of the most marginalized and isolated farmers into the research and decision making process.

Even though the local and unwritten knowledge of farmers is usually difficult to articulate and, above all, to be institutionalized and introduced into the policy strategies, the project aimed at bringing local knowledge of farmers directly into the decision making process. This integration allowed dealing with the services to be provided in more flexible way, since many times the technical approach that resulted was too rigorous while the local knowledge was crucial to enhance the effectiveness of the services provided.

Moreover, this collective action made use of the local knowledge of farmers but also pushed farmers to increase their knowledge of the territory and, above all, allowed for the transfer the new knowledge generated into the public administration, with the integration of new data and new information that was considered strategic to plan the activities related to flooding prevention.

Indeed, another aspect related to the knowledge generation is the contribution of local farmers to the local and regional cartography, since in many cases they have registered canals and streams with names, location and status of maintenance. Thus, the project allowed the local agency to integrate the information collected by farmers into the Regional Geographic Information System (Fig. 4.6).

Table 4.2 The actors involved in the "Custody of the Territory" project

EU and national level	*Role*
European Commission	Approval and design of Rural Development Policies
Ministry of Agriculture	Legislation on Multifunctional Agriculture and National Strategic Plan for RD
Regional government	Design and management of RDP
District level	*Role*
Pistoia and Lucca Provincial authorities	Local implementation of RDP (LRDP)
Mountain Community Media Valle del Serchio	Setting and managing the PES scheme
Other Mountain Communities	Technical assistance
Farmers' organizations	Administrative assistance
Local level	*Role*
Municipalities	Results dissemination, direct contact with farmers and technicians
Farmers, technicians, local communities	Environmental services delivery (implementation rules, social norms)

At the same this co-produced knowledge was particularly useful in planning the intervention works to be carried out in the district. The first-hand knowledge that the farmers acquired through the monitoring was particularly important in updating—and in some case correcting—the geographical data base of the local agency.

4.3 Institutional Arrangements

4.3.1 The Actors Involved

As discussed in Chap. 2, a very relevant issue to address when assessing the dynamics of collective action in the context of public goods provision is what type of organization has addressed the agro-environmental problems. Several studies show the differences, in terms of efficiency, success and outcomes, according to the type of organization involved, for example between an organization controlled directly by farmers of controlled by a national/regional government. As already demonstrated, the main authority involved in the project is the Mountain Community 'Media Valle del Serchio', the local authority which defined the contracts, coordinated the project and maintained the information system.

At the same time it is possible to identify the involvement of various institutions and organizations that, to a various extent, have provided the institutional, technical and policy support to the project. Table 4.2 summarizes the main actors involved and their role in the project.

It is possible to recognize different levels of involvements in the project. The first level is related to international (European Commission) and national (Ministry of Agriculture, Regional Government of Tuscany) levels, that are indirectly involved through the development of the necessary supporting policies (rural development policies, agro-environmental measures) and through the national and regional legislative framework enhancing the multifunctional role of agriculture.

The project was developed thanks to the national and regional legislative framework on multifunctional agriculture that has been developed during the last decade. A very important step is the 2001 national Legislative Decree n. 228, which redefined the role of farmers, through an acknowledgment of the environmental services that farmers may provide in rural areas. This decree (art. 15) allows 'local institutions to draw up agreements with farmers in order to facilitate ecosystem services, the preservation of forest and agricultural landscape, the hydro-geological management of the land'.

The representatives of local institutions involved in the project have emphasized the importance of this legal instrument, since the legal acknowledgment of the multifunctionality of agriculture has allowed local government authorities to set different types of agreements with the farmers. Indeed, the legislative Decree n. 228 allowed for the broadening of the spectrum of activities and services that farmers could provide, translating the academic and institutional debate on multifunctionality into agreements that before could not rely on a coherent legislative framework. Nevertheless, the representative of the local authority argued that there were several important issues that were not yet sufficiently regulated by the existing legislation. This is the case of the monitoring activities: it is not clear whether the law includes this 'immaterial service', which is difficult to quantify in terms of economic return for the institutions (and for the society) in terms of costs for farmers. Despite the difficulties in regulating and supporting this service, the local authority decided to include it into the PES schemes and this is unanimously considered the most innovative aspect of the project. Indeed, this control of the territory carried out by the farmers allowed for the developing of a knowledge-based flooding prevention strategy, since the monitoring enabled the identification of the areas prone to particular risks and it was an important tool for the local authority in planning its activities. According to the local stakeholders, the national legislation on multifunctional agriculture was interpreted by the local agency in very innovative way, and this resulted in a more comprehensive scheme, with a payment addressed not only to physical goods, but also to the prevention activities, primarily through the monitoring.

According to the local stakeholders, the regional RDP should also include the specific support for monitoring activities, since to date the project managers have failed to find a set of RDP measures which could finance the project in its entirety:

> In the regional RDP we could not find a set of measures that could exactly fit for the project. The RDP philosophy is based on investments, works and interventions that have a beginning and an end. Our project is based on prevention, on continuous monitoring activities and in small interventions that are not possible to finance through the inflexible instruments of the RDP (INTERVIEW n. 4).

From this perspective, the project coordinator has emphasized how this approach, namely favoring a collective action through the involvement of farmers living in less accessible areas, has been a strong political choice which has brought about several complaints, especially from the specialized cooperative and firms that traditionally carried out these types of environmental services. At the same time, the project coordinator recognized that this choice was crucial in achieving the primary objective of the project, which is acknowledging, through the institutional and the economic support, the added value of local farmers in providing environmental goods and services in the areas surrounding their farms.

With regard to the support to the PES into the Rural Development Policies, the case study shows that the effective implementation of local strategies also depends on the development of coherent supporting policies. The regional implementation of the RDP in Italy results in a very broad differentiation at the national level, since every Regional government set a Regional Development Programme and a Regional implementation document. Local programming is articulated on the basis of the Regional implementation document and it is based on two types of strategies: Local Rural Development Plans (LRDP), which are provincial programmes, and Integrated Local Development Strategies (ILDS), which are the programming activities related to the Local Action Groups (LAGs).

With regard to the financing support to the activities related to the project, the role of Provincial authorities is particularly relevant, since Provincial governments in Tuscany are in charge of the local implementation of the rural development policies (through Local Rural Development Plans, LRDP). At the same time this devolution may lead to differentiated results according to the different priorities of the provincial administrations. This is the case with the provinces involved in the project 'Custody of the Territory', which have set different priorities for the implementation of the LRDP: a higher amount of funding for the hydro-geological activities in Pistoia province compared to Lucca province did not ensure an homogenous effect of the project on the district.

Nevertheless, according to the stakeholders interviewed, in both provinces the municipalities have played a significant role in maximizing the effectiveness of the project, by disseminating the results and by encouraging direct contact between the technicians working for the municipalities and the farmers; and in some cases, by also involving directly the 'farmer custodians' into the early intervention works. At the same time, the project coordinators agree that a greater involvement of the municipalities could provide a further improvement in the management of the project, since this could result in a greater involvement of local communities.

Among the actors who have played an important role in the project, respondents also cited the farmers' organizations, which in many cases have played a very important role in motivating farmers to participate into the project, by explaining to them the innovative approach of this initiative as well as the opportunity to diversify their income and to improve the specialization towards the environmental services as a strategic investment for the future. Moreover, during the early stages of the project, farmers' organizations also supported the farmers for the administrative tasks of the initiative, and in some cases gave advice and information to the local agency during the selection of farmers, thus providing some information about their reliability and their professionalism.

The analysis of the institutional arrangements of the project also provided some insights into the management of rural development policies at the local level, especially with regard to coordination mechanisms needed for a more effective provision of environmental services through collective action. One issue that all representatives of the local agencies have highlighted is the lack of coordination amongst the local institutions which, at different levels, are involved in the hydro-geological management of the territory.

The local representatives interviewed have emphasized how the landscape management activities would need a coherent and homogenous approach at river basin level, since the territorial needs often do not coincide with the administrative boundaries (regions and provinces). While the hydro-geological management of the territory and the resilience to flooding are highly dependent to the correct management of rivers, canals and streams, the environmental priorities are usually addressed on the basis of the different administrative levels (municipalities, provinces and regions). In some cases inflexible policy tools and institutional arrangements based on the administrative borders hindered a more effective approach to the broad range of environmental services to be provided at landscape level.

While the success of initiatives such as 'Custody of the Territory' is closely related to the coordination of local institutions and local agencies, especially in terms of management of duties amongst river basin authorities, municipalities, mountain communities, provinces and municipalities unions, according to the respondents a more effective coordination amongst these authorities would have favoured a more efficient support for this collective action, by providing a wider spectrum of services carried out by local farmers, such as maintenance of mountain paths, fire services prevention and other environmental services.

4.3.2 Information, Communication and Joint Learning

The role of communication and information regarding agro-environmental strategies is usually an issue that economists fail to address or do not address adequately, but it plays a very crucial role in determining their success or failure, and this is particularly true where grass roots collective actions are involved (Ayer 1997).

From this perspective, it is interesting to explore the use of the new communications technologies in 'Custody of the Territory' project, where the development of specific technological tools was shown to have a strong potential in coordinating the different stakeholders, with the objective of increasing the effectiveness of the environmental services to be provided.

During the second phase of the project the responsible parties of the local authority decided to standardize the procedures for the monitoring activities (report and pictures provided by the farmers) with the main objective of garnering all this information into a single database. Thus, an information system based on *google maps* was created, named IDRAMAP. This is a web site that is possible to access

Fig. 4.7 IDRAMAP. (Source: Comprensorio di Bonifica Media Valle del Serchio)

from the home page of the local agency, and it was developed as an on-line information system. This system was also developed with the objective of expanding the monitoring activities to the local inhabitants.

Local farmers and citizens, by accessing the IDRAMAP, can signal the need for intervention on a specific location of the area simply by clicking on the map. When clicking on the corresponding point of the map, IDRAMAP opens a menu that allows the user to select the type of environmental problem detected.

According to those responsible for the project, the role of this on-line information system is twofold. On the one hand, by extending the monitoring activities to the local population, it represents a very effective tool for communicating critical situations and problems of various kinds; on the other hand, IDRAMAP was developed to make public the activities carried out by the local agency, in order to increase the transparency of the use of revenues generated from local reclamation tax. However, those responsible for the project have pointed out that this increased visibility may also expose the local authority to some risks, since its activities may be checked out by the local citizens, who can challenge the local agency in case of delay of intervention and in case of the high number of warnings not corresponding to an adequate number of intervention works (Fig. 4.7).

Nevertheless, the coordinator of this information system recognized that IDRAMAP is an important management tool for the agency, since it allows us to rationalize the activities and to reduce the costs by avoiding double warnings, by planning joint inspections and maintenance activities in specific areas.

From the point of view of the procedure, IDRAMAP is structured in four steps: (1) warning (2) acceptance (3) attribution to the institution in charge (4) acceptance (Fig. 4.8).

Fig. 4.8 The four steps of IDRAMAP. (Source: Comprensorio di Bonifica Media Valle del Serchio)

One of the most critical aspects of IDRAMAP is related to step (3), namely the attribution to the institution in charge. Indeed, according to the representative of the Mountain Community 'Media Valle del Serchio', the lack of coordination amongst local agencies and local institutions described above do not allow us to make full use of IDRAMAP, which could potentially assign the type of intervention signalled directly to the local authority in charge to solve the problem. A better coordination amongst the various institutions could lead to higher recognition and use of IDRAMAP by the all the local agencies involved in the management of the territory at district level (river authorities, provincial authorities, municipalities, etc.).

Nevertheless, the development of this innovative learning/information system, based on co-production, emphasizes the importance of negotiation amongst the different local stakeholders and, above all, the importance of joint learning processes and knowledge generation. Indeed, an approach to PES which stimulates the formal and informal interaction amongst several rural stakeholders may facilitate the increase of social capital and institutional capital, by increasing the co-operation and coordination at territorial level.

From this perspective, it may be argued that effective PES are based on a knowledge system as well as a system of rules, but also on an effective information and communication system which codifies that knowledge and those rules and favours the interactions amongst the local stakeholders.

4.4 Benefits, Barriers and Potentialities of the Scheme

4.4.1 Factors Affecting the Collective Action

As described in sect. 2.1, the main factors that influence collective action may be categorized into four groups: (1) resource system characteristics, (2) group characteristics, (3) institutional arrangement and (4) external environment.

Based on this classification, Table 4.3 summarizes the key successful factors in the 'Custody of the territory' project.

In terms of the characteristics of the resource system, in the area involved, the risk of hydro-geological disorders is very high, and farmers were aware that a lack of correct management of rivers and riverbanks could have important negative effects

Table 4.3 Factors affecting collective action in 'Custody of the territory' project

1) Resource system characteristics	*2) Group characteristics*
Lack of hydro-geological management	Social capital: trust and mutuality
Lack landscape management due to abandonment of mountain areas	Involvement of marginal and isolated farmers
High risk of flooding, especially in urbanized areas of the plains	Participatory events: involvement of local community
	Interdependences local agency-farmers
3) Institutional arrangements	*4) External environment*
Simple rules	Payments for Environmental Services settled by Mountain Community
Different rural stakeholders involved	National and Regional legislation on multifunctional agriculture
Information/early warning system: IDRAMAP	Measure 226 of the Regional RDP
Co-production of knowledge: joint inspections	

(such flooding and bad landscape) for their business. At the same time the single farmers were not able to implement individual solutions to this problem, and a reduction of the flooding risk was clearly the most important joint product of this collective action. At the same time more immaterial joint outputs were produced, such as an increasing sense of community amongst the farmers of the same areas, an acknowledgment (even though not at the same level everywhere) by the local community of the important social and common functions carried out by the farmers.

One of the instigators of the project synthesized well this innovation in the following statement:

> The project was developed to contribute to the prevention of hydro-geological problems and to increase the resilience of flooding in the area through continuous monitoring and early intervention of local farmers. This is a completely different philosophy compared to the traditional PES, since here the main issue is recognizing, developing and financing all those small preventing actions and the first maintenance works which local farmers carry out in those territories (INTERVIEW n. 5).

As described above, the approach adopted in this collective action was mainly developed by the local authority Mountain Community 'Media Valle del Serchio', which set a payment for environmental services aimed at providing flooding prevention, mainly through information and an early warning system based on local communities and local farmers, who can identify and tackle problems before the environmental disasters such a flooding take place. The information system also aimed at building social networks that could enhance the formal and informal exchanges amongst local institutions, advisors and farmers, with the main objective of increasing the efficiency of the environmental services provided.

Thus, the project, together with the provisions of the environmental public goods such as the 'hydro-geological maintenance' and the 'resilience to flooding' has aimed at increasing the institutional and social capital through the provision of relational goods, which may be considered the real innovation of this initiative.

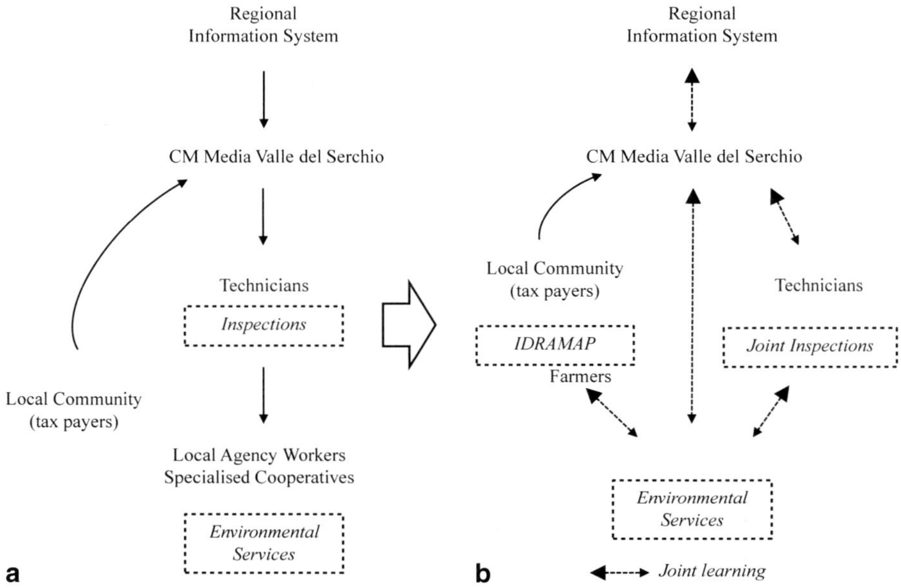

Fig. 4.9 The development of a new information/learning system. **a** The hierarchical approach in place before. **b** The information/learning system developed through the project

This set of objectives was pursued by encouraging the direct involvement of farmers into the project, by selecting and training them not only on the basis of their technical capacity (ownership of machineries) but also on the basis of their location and on their knowledge of the territory. From this perspective, farmers interviewed have emphasized that they have strongly appreciated this approach and that it makes them feel that the main actors are engaged with the management and maintenance of the area that they monitor.

This system, which aims at creating and spreading awareness among the local actors on the role that farmers may play in the delivering of important public goods, favored the co-production of environmental services by the local agency, the advisory system and the local communities (Fig. 4.9).

Thus, the co-production approach on PES emphasizes the importance of negotiation amongst the different local stakeholders and, above all, the importance of joint learning processes and knowledge generation. The case study also shows that collective action led by local agencies may stimulate the formal and informal interaction amongst several rural stakeholders and that subsequently this may facilitate the increase of social capital and institutional capital, by increasing the co-operation and coordination between different rural stakeholders. The payments and the support received by farmers are not only an economic incentive to deliver environmental services, but are structured as an incentive to actively participate in the environmental management of the territory, by increasing the relations and interdependencies amongst farmers, local institutions, advisory systems and local communities.

The representatives of the local institutions interviewed (local agency, provinces, municipalities) argued that an important strength of the project is its simplicity: both in terms of measurement design and in terms of their implementation. Indeed, the management of this collective action is based on a daily relationship between the coordinator of the project, the technicians and the farmers. This strong collaboration has favored the development of trust and a willingness to cooperate and has facilitated the implementation of a very simple agreement without excessive regulation or bureaucratic tape. This simplicity was stressed especially in contrast to the RDP measures, which were described as 'cumbersome and rigid', since these measures usually required a deep knowledge of the administrative procedures that in many cases discouraged their adoption by the farmers. On the other hand, the agreement experienced in the project shows how a direct relationship between a local authority and the farmers may facilitate the adhesion to the collective action, also by increasing its effectiveness.

4.4.2 Benefits and Barriers of Collective Action

Through the analysis of the case study it was possible to identify a series of opportunities for the project, especially in terms of increasing the range of environmental services to be provided but also in terms of scaling up and building new experiences by using the same approach.

In general terms, the representatives of the local agency have stressed their satisfaction with the effectiveness of this collective action considered strategic for their activities, even from an economic standpoint, with an undoubted saving in terms of labor, equipment and monitoring activities. This was confirmed by the fact that all respondents concurred that one the main strengths of the project was its cost-effectiveness.

With regard to the specific innovation of this strategy, the project coordinator was also highly satisfied with the work carried out by the farmers, since they have been shown to be suitable actors to carry out the required hydro-geological management activities.

The main strengths of this collective action, according to the farmers interviewed, are related to the creation of the new identity of local farmers as 'custodians', which implies a strong commitment to the environmental management of the area. The most pro-active farmers are becoming points of reference for local people and, above all, they represent a reliable information network and an efficient early warning system for the public administrations in charge of the environmental management of the district.

The main opportunities of the project also concern the increase of their knowledge of the areas surrounding their farms as well an increase of their reputation within the local community. This also increased their visibility and for some farmers new contacts with other farmers have been established as result of this project.

The analysis of the institutional arrangements of the initiative may also provide some insights into the management of rural development policies at the local level, especially with regard to coordination mechanisms, which can result in a more effective provision of environmental services. Indeed, one issue the majority of the stakeholders interviewed have highlighted is the lack of coordination amongst the local institutions that, at different levels, are involved in the hydro-geological management of the territory.

Moreover, the responsible parties of the local authority highlighted how the effectiveness of this collective action, both in terms of environmental services provided and in terms of involvement of local farmers is not homogenous on the territory.

The project coordinator pointed out how the initiative is more efficient when the environmental services are provided by the more structured and larger farms, while the small farmers often demonstrate some difficulties in terms of simple procedures and administrative tasks of the project. To some extent these difficulties are due to the lack of experience of small scale farms in working for the public administration and, more in general, the lack of experience for works that required different administrative tasks from those that they are used to undertaking. The administrative burdens of the project are mainly related to the different ownerships of the land, which require different types of administrative tasks (decrees, orders, and specific authorizations).

Another barrier to the further development of the initiative is the planning of interventions, since the coordinator before assigning the PES to farmers must take into account the availability of machineries and the competencies to carry out the works. While the larger farms usually have the necessary equipment to perform almost all operations, in some cases, small farmers may overestimate their ability to intervene, both in terms of equipment and of technical knowledge.

The difficult task of the managing body is balancing the need for improving the management of the project, focusing on the farmers that have been shown to be able to cope with the task required and maintain its main objective, by selecting local farmers according to their location but also according to the effectiveness their environmental action.

Finally, the most critical element of the project lies in the difficulty of scaling up and becoming the dominant approach of PES delivery in the district. As pointed out by a representative of the Mountain Community Media Valle del Serchio:

> Our local authority should increase the visibility of the project at regional level, in order to increase its scope, broadening its action and obtaining specific funding; on the opposite, so far this project has failed to expand and to link to other similar initiatives and actually it may can be considered simply an 'experimental proof' that adds to the other initiatives related to the reclamation activities (INTERVIEW n. 1).

Indeed, the financial gains of the project could have resulted for example, in a reduction in the local tax for reclamation, and this could have increased the awareness of the local population on the effectiveness of this collective action. At the same time it is clear that in order to increase the scope of the project it would be necessary to include it in the regional agri-environmental strategies, in order to experiment similar actions in other regional districts also through the development and implementation

Table 4.4 Benefits and barriers of collective action in 'Custodians of the Territory'

Benefits	Barriers
Cost effectiveness (saving in terms of labor, equipment and monitoring activities)	Lack of coordination amongst local institutions
Creation of a new identity for 'farmers custodians'	Environmental services to be provided do not correspond to the administrative boundaries
Increasing commitment of farmers for the environmental management of the area	Lack of equipment and of technical knowledge (especially for small farmers)
Knowledge generation and learning	Lack of specific training
Institutional capital and capacity-building	Difficulties in scaling up and expanding the collective action
Increased reputation of farmers	
Building up of a network of local farmers	

of specific regulations and of specific financial resources for collective actions aimed at flooding prevention. This is hindered by several political and administrative barriers, but also to the financial mechanisms for the environmental action. As put by a promoter of the initiative:

> Large financial resources for mountain areas are available only in cases of environmental disasters such as flooding; and it is evident that in many cases the public and private interests are primarily focused in intercepting these funds; the approach experimented in this project is completely revolutionary because is based on prevention rather than on maintenance. This action was not developed to obtain funding, because it does not need much funding; for this reason local politicians and local entrepreneurs are not very interested, because it is based on a conception of the environment is not business oriented (INTERVIEW n. 1).

Finally, as has already been described, one of the main barriers of the project is related to the lack of coordination between local institutions (river basin authorities, province, region, CM, etc.) that so far has prevented the scaling up of the project, that could be adopted at a larger scale (Table 4.4).

4.4.3 An Innovative Model to Deliver Environmental Services

The project 'Custody of the Territory' was created in order to provide and protect a large set of environmental goods and services closely linked to agricultural activities, such as landscape, soil protection, resilience to flooding, but also non-environmental public goods, such as social capital, institutional capital and new knowledge. This approach was promoted by a local authority which was able to involve local farmers in the delivering of environmental services through a re-organization of institutional and social settings that facilitated co-production.

This collective approach is quite innovative. Indeed, traditionally PES are settled on a simple mechanism, based on a voluntary participation of farmers, where the payments are conditional to the environmental services effectively provided: farmers provide environmental services and get paid for doing so ('provider gets') while those who benefit from environmental services pay for their provision ('user pays') (see box 5.1).

As shown by Engel and Palmer (2008), the efficiency of the PES is related to the following conditions (i) clear relationships between the type of land use being promoted and the provision of the environmental service to be provided; ii) the possibility for the stakeholders to strictly follow the rules of the contracts and (iii) the presence of an efficient monitoring system, which must ensure that the services are really provided. As discussed by Muradian et al. (2010), in reality it is very difficult for the PES to comply strictly with these conditions. On the other hand, when implementing PES, it is usually necessary to take into consideration the collective dimension and the coordination mechanisms, with the main objective of changing individual and collective behaviors that otherwise would lead to excessive deterioration of ecosystems and natural resources.

As highlighted by Vatn (2010), in designing PES, it is really important to distinguish incentives and compensations, since these different approaches involve different relationships between the involved agents.

The case of 'Custody of the Territory' shows how the local agency has aimed at creating not an instrumental relationship based on compensation, but a more complex system of incentives, rules and knowledge, which is based on reciprocal relationships, trust and engagement. Thus, the co-production mechanisms observed in the case study involves a more complex relationship between 'provider gets' and 'user pays', with a more integrated strategy based on the development of a local network of farmers, citizens, advisory system and local institutions.

Through co-production, the PES received by farmers are not only an economic compensation for delivering specific services, but are structured as a set of incentives to push farmers to actively participate in the environmental management of the territory, by increasing their relations and interdependences with local institutions, advisory systems and local communities.

The case study shows that the local authority in charge of coordinating the project has not only stabilised the PES, but has also provided opportunities for personal growth and learning for the advisors and the farmers involved. The learning opportunities have been based on two types of strategies, one addressed to the broad involvement of the local communities and the other on the exchange between professionals (technicians) and farmers. In the first case a specific website (IDRAMAP) was created in order to implement an early warning system and to rely on a common data base. This tool allows the blurring of the distinction between services providers and service users, facilitating co-production of knowledge. Indeed, IDRAMAP was implemented to reconfigure the ways in which environmental services were developed and delivered: local communities are not only are the service users but, by signalling the environmental problems are, to some extent, also service providers (Fig. 4.10).

The distinction between service providers and service users were reduced even more in the case of farmers, since the local authority offered to farmers a range of incentives in order to transfer their knowledge and capabilities into its strategy. The Mountain Community Media Valle del Serchio then became a catalyst and a facilitator for the development of a network of local farmers, by encouraging co-production of knowledge, reciprocity and mutuality amongst local actors, who were enabled to provide environmental services in a collective manner.

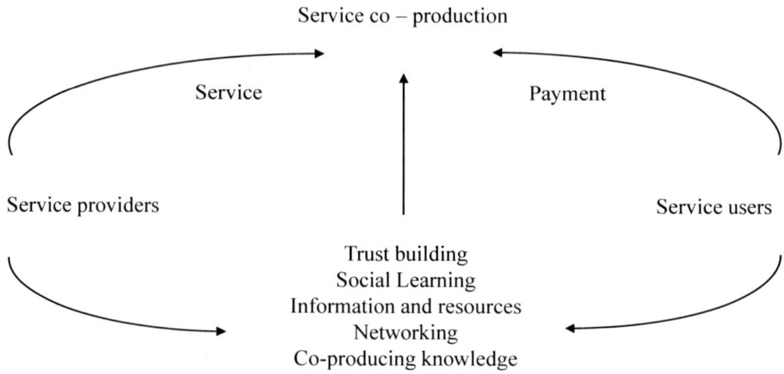

Fig. 4.10 A new model of PES, based on co-production

It may be argued that the application of the concept of co-production on PES implies a different approach to the jointness amongst agricultural activities and the services provided by farmers. While the neo-classical approach focuses mainly on the remuneration of the externalities and of the countryside goods provided by farmers, a co-production approach broadens the scope of jointness by pointing to the ways to reinvigorate farmers' motivations and attitudes and to rebuild their knowledge in order to realize their full potential in delivering environmental services. This broad conceptualization of PES implies a greater involvement of farmers in the decision making processes, by encouraging them to use their human and professional skills to deliver the required services.

To summarize, this collective action shows the potential for developing a new management model of environmental services, which allows for the maximizing of the effectiveness of control activities on the territory, by reducing the flooding risks through prevention and through early intervention works ensured by a network of local farmers.

References

Ayer HW (1997) Grass roots collective action: agricultural opportunities. J Agric Resource Econ 22(1):1–11

CNG (2010) Rapporto sullo stato del territorio italiano. Centro studi del Consiglio nazionale dei Geologi (Cng) e Cresme

Corburn J (2007) Community knowledge in environmental health science: co-producing policy expertise. Environ Sci Pol 10(2):150–161

Cooper T, Hart K, Baldock D (2009) The provision of public goods through agriculture in the european union. Institute for European Environmental Policy, London

EEA (2008) CLIM17 river floods. European Environment Agency, Copenhagen

Engel S, Palmer C (2008) Payments for environmental services as an alternative to logging under weak property rights: the case of Indonesia. Ecolog Econ 65(4):799–809

Legambiente (2011) Ecosistema Rischio 2011. Monitoraggio sulle attivita delle amministrazioni comunali per la mitigazione del rischio idrogeologico. Legambiente, Roma

LUC (Land Use Consultants) (2009) Adapting agricultural policy to increased flood risk. Land Use Policy Group, UK

Muradian R, Corbera E, Pascual U, Kosoy N, May PH (2010) Reconciling theory and practice: an alternative conceptual framework for understanding payments for environmental services. Ecolog Econ 69(6):164–189

Vatn A (2010) An institutional analysis of payments for environmental services. Ecolog Econ 69(6):1245–1252

Chapter 5
The Agri-environmental Agreement in Valdaso (Marche)

Abstract Chapter 5 describes the innovative and collective approach to agri-environmental action which was experienced in Valdaso area (Marche region), where a group of farmers started a grass roots initiative to adopt integrated agriculture at territorial scale with the objective of protecting water and soils from pesticide and nitrate pollution. This collective action was supported by the regional and provincial authorities, which settled a territorial agri-environmental agreement (AEA) financed by the regional Rural Development Programme. The case study shows that farmers' collective action may play a significant role in controlling negative externalities from agriculture, especially if local institutions positively influence collective decision making behaviors, by structuring a range of incentives, capacity building programs and technical assistance to align individual and group interests and, above all, private and public goods objectives. The last section of the chapter explores the ways in which this collective approach to agri-environmental action could be better embedded and institutionalized in the current policy settings.

Keywords Agri-environmental agreement · Integrated agriculture · Co-management · Rural Development Programme

5.1 Case Study Overview

In the Valdaso area (Marche region), a group of local farmers specializing in fruit production started a grass roots initiative to adopt integrated management techniques at a territorial scale, with the objective of protecting water and soils from pesticide and nitrate pollution. This initiative has been supported by the regional and provincial governments, which set up a territorial agri-environmental agreement (AEA), mainly financed by the regional RDP.

An agri-environmental agreement (AEA) may be defined as 'a set of commitments for farmers in a limited area, supported through a mix of RDP measures, that can be activated to reach specific environmental goals. Based on a territorial approach and by involving public and private actors in the context of a shared project, AEAs are aimed at implementing collective and coordinated actions for the management and improvement of the environment' (Marche Region 2007).

F. Vanni, *Agriculture and Public Goods*,
DOI 10.1007/978-94-007-7457-5_5,

The AEA implemented in the area is part of a wider project, called the 'Valdaso project', which is an integrated project involving different actions related to the rural vitality of the area, the renovation of the small town centres and the cleaning and management of the Aso river. This project is related to the sustainability of the area and it is an attempt to harmonize the different policy tools for the sustainable development of Aso Valley, by integrating the rural development strategies with territorial planning and with the local policies implemented for management of natural resources.

The area covered by the Valdaso AEA is the Aso Valley, a territory alongside the boundary between Ascoli Piceno and Fermo Provinces, which follows the path of the Aso river. Valdaso is a well preserved valley with a very attractive landscape, but the environmental quality of the river has suffered from various types of anthropogenic pressures, especially derived from intensive agricultural production.

Indeed, the local agriculture is highly specialized in fruit production (peaches, plums, apples and pears) and the valley concentrates almost 60 % of the regional production and processing industries in the fruit sector. The orchards in this valley have been traditionally cultivated with a high use of chemical inputs (fertilizers and pesticides), with negative consequences on the river's water quality, on agricultural soils and also on the operators' health. Nevertheless, in this area, during the last decade, an increasing awareness has been observed, among the local communities and the local farmers, of the negative impacts of this intensive and highly specialized type of farming. Thus, in 2007, a small group of farmers (allied in a local farmers association *Nuova Agricoltura*—'*New Agriculture*') started a grass roots initiative to adopt integrated management techniques at a territorial scale. This initiative has been supported by the regional and provincial governments, which settled a specific agri-environmental agreement, financed by the regional RDP.

The agreement established specific targets, to be achieved in a period from 5 to 7 years, such as the reduction of 30 % in macronutrients (nitrogen, phosphorus and potassium) used in the territory and the substitution of agri-chemical inputs, characterized by acute or chronic toxicity, by 90 and 85 % respectively. To achieve these results, the AEA was structured as an integrated package of measures of the regional RDP, with the aim of financing a set of initiatives that could support the adoption of more sustainable agricultural practices at territorial level. The package of measures comprises the measure 111 on training activities and information actions and the following sub-measures of the agro-environmental schemes (214): integrated pest management (IPM) techniques, mating disruption, organic farming and protection and improvement of soil through green cover.

Together with the actions specifically implemented to reach environmental objectives (measure 214), the measure 111 was included in the agreement with the objective of raising farmers' awareness on the impacts of their farming practices on the environment, as well as on the role of farming in protecting the environment and enhancing the rural landscape. Indeed, through this RDP measure, a capacity building programme for farmers was established, with specific training regarding the technical guidelines on integrated agriculture, articulated in farm visits and specific workshops, which were organized to increase information sharing among local farmers regarding the environmental, economic and health effects of IPM techniques.

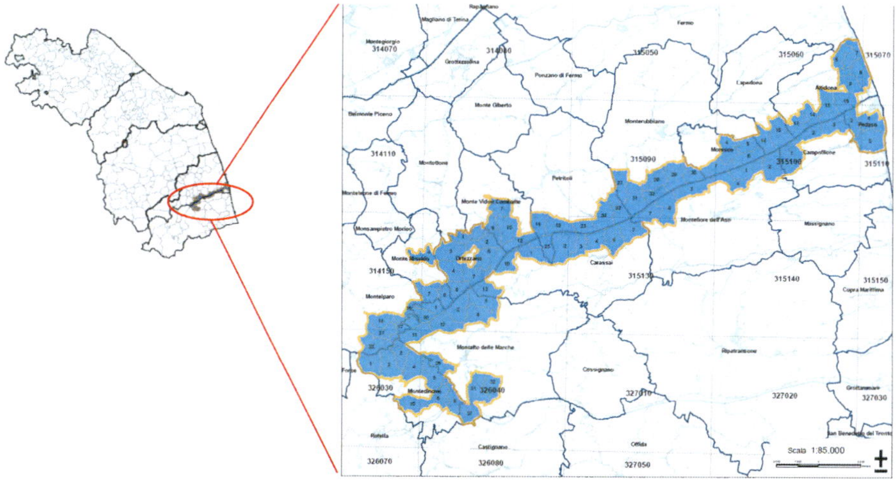

Fig. 5.1 The designed area of Valdaso AEA. (Source: Ascoli Piceno Province (2009))

As prescribed by the regional RDP, the area covered by the AEA should be an unbroken piece of land belonging, for more than 50 %, to the Nitrate Vulnerable Zones (NVZ). The total area under the AEA should be at least 1,000 ha and the UAA cultivated with fruit trees (peach, plum, apple and pear) must represent at least 5 % of this area (Marche Region, 2008). The designated area of Valdaso AEA encompasses 50.4 % of NVZ areas (Fig. 5.1).

During the first year of the agreement (year 2009) 82 farms were involved, corresponding to 257 ha cultivated with Integrated Pest Management (IPM) techniques. In the following year (2010), other farmers joined the AEA, but in a smaller number, also because a condition for access to the agreement is that areas involved should be in a NVZ. In the last tender (2011), however, a further 14 farmers have applied to join the Valdaso AEA, and at the beginning of 2012 about 100 farmers were involved in the agreement.

5.2 The Territorial Agri-environmental Agreement

5.2.1 The Development of a Territorial Agreement

According to the local stakeholders, Valdaso AEA was implemented to reduce the environmental and social risks related to the high use of chemical inputs for fruit production. Farmers particularly have emphasized how they have increasingly perceived the need of minimizing the negative impact of local agriculture on river water, on air quality, on soil functionality and, above all, on people's health and on food safety. As put by a farmer involved in the agreement:

> Historically the main environmental problem of the area is the very high use of pesticides, and maybe we have realized too late that we have heavily altered the ecosystem; in my opinion it is not so easy to get back to the right equilibrium… I remember that 20 years ago during the winter the soil in the valley was almost yellow and white and it smelled badly, due to the aggressive use of pesticides used by local farmers (INTERVIEW n. 28).

Thus, the most motivated and younger farmers created *Nuova Agricoltura*, a spontaneous and informal association (which gathers no more than 20 farmers), where the associates share a common view on the need for adopting more sustainable farming practices at territorial level, in order to reduce the negative environmental impacts described above.

The birth of the *Nuova Agricultura* dates back to 1995 when there were plans to build a gas plant and the passage of a pipeline in the valley. The farmers joined this association in order to increase their lobbying power and to organize some information campaigns as they were particularly concerned about the expropriations of land but also for environmental and landscape impacts of these new infrastructures in the area.

Nuova Agricoltura was also created in order to address in a more effective way the technical, economic and social issues related to the transition towards sustainable agriculture that, according the farmers interviewed, official farmers' organizations failed to address. Indeed, according to the local farmers, the success of this association is based on its informal character, since it is based on the direct knowledge of people, on relationships of friendship, reciprocity and trust. As put by one of the pioneers of *Nuova Agricoltura*:

> We used to meet up at the café every night; everything was born there, and from there we started to organize some meetings at the old school of the town in Montalto, where we discuss about our projects and where we share our opinions and ideas. However, we are also in touch by telephone, and we meet often, since we all live in this valley. A few of us who have become very close friends, we trust each other (INTERVIEW n. 25).

Farmers believed that an informal association could have important positive effects, since the aggregation of farmers could potentially have a higher 'lobbying power' than individuals, by increasing the credibility and legitimacy of the joint action to be undertaken.

According to the local stakeholders, the development and implementation of the AEA has been possible because of the activities of this informal association, but also as result of the experiences that almost all local farmers have had in the past as associates to a cooperative. The technician responsible for the AEA (project leader) working for ASSAM (*Agenzia Servizi Settore Agroalimentare delle Marche*, the regional agricultural advice agency) has emphasized both the positive and negative impacts of this cooperative for the new project as follows:

> Putting them together was not difficult because I knew them all, we started to have meetings; in the past they were already associated in a cooperative and this helped a lot. I have started my professional career in this cooperative 20 years ago and this cooperative was the first one adopting integrated pest management production in the area. At the same time the cooperative went bankrupt, even though farmers recognized the need to work together, at the beginning they were scared because of this negative past experience (INTERVIEW n. 14).

Indeed, even though the previous experience in this cooperative has been positive from the point of view of the habit of local farmers to collaborate, for some of them the bankruptcy of previous cooperatives has been a major barrier to a new type of aggregation. *Nuova Agricoltura* played a crucial role in creating the conditions to aggregate the farmers and to convince the most hesitant ones to work together again. This association was also created in order to promote exchanges and collaborations for many initiatives such as the AEA but also additional projects in the future, for example regarding the joint supply and marketing of local products.

With regard to the development of the AEA, farmers who decided to create the association were already testing IPM techniques in their farms and they realized that this type of technique would have implied some level of farm aggregation to be more effective; especially mating disruption, in order to be effective, requires large areas treated with pheromones and, since in the Marche Region the average size of farms is quite small (10.2 ha), a collective approach was needed to make the technique successful.

The experimental phase with the IPM techniques began in 2007 with a small group of farmers associated with *Nuova Agricoltura*, but the agreement was made in 2009. During these 2 years the most motivated farmers, partly as a result of the encouraging results obtained in their orchards, succeeded to involve a very large number of local farmers (almost eighty). The farmers who already experienced these techniques were particularly motivated to involve other farmers since, as mating disruption is effective only whether a large piece of unbroken land is involved. The high number of farmers involved is the result of the effectiveness of the techniques proposed but also to the power of persuasion of farmers from *Nuova Agricoltura*.

To increase the land affected by mating disruption, some farmers have also installed dispensers in the orchards of the most reluctant neighbours, in order to cover a larger area and thus maximize the efficiency of the IPM. As stated by this farmer:

> I started to use the IPM 2 years ago; the first year went so well, but the second year it did not work so well. The ASSAM technicians and I realized that one of the main problems was the presence of the insects on the nuts tree around my farm. So I bought the dispensers to be installed in these nut trees around my farm. Fortunately the owner was happy ... of course, I have protected his trees at my expenses .... But fortunately it was a good idea, because this year the technique worked well. (INTERVIEW n. 28).

All respondents argued that one of the most interesting and innovative aspects of the AEA was to stimulate and test this collective approach, where farmers were concentrated to the territorial effects of this agro-environmental strategy, also as result of the characteristics of the new farming practices adopted. These aspects have been well synthesized by the ASSAM project leader:

> To obtain important results in the field, farmers must change the mentality, and to obtain this institutions and advisors must also stop thinking in terms of a single measure, it is necessary to have a project/integrated approach, in order to involve people in more innovative things. Farmers are usually considered ignorant, but if you will involve them you realize that they can give you impressive feedback. They are not ignorant, but isolated. It is then necessary to involve them and to make use of their unbelievable enthusiasm to act collectively (INTERVIEW n. 14).

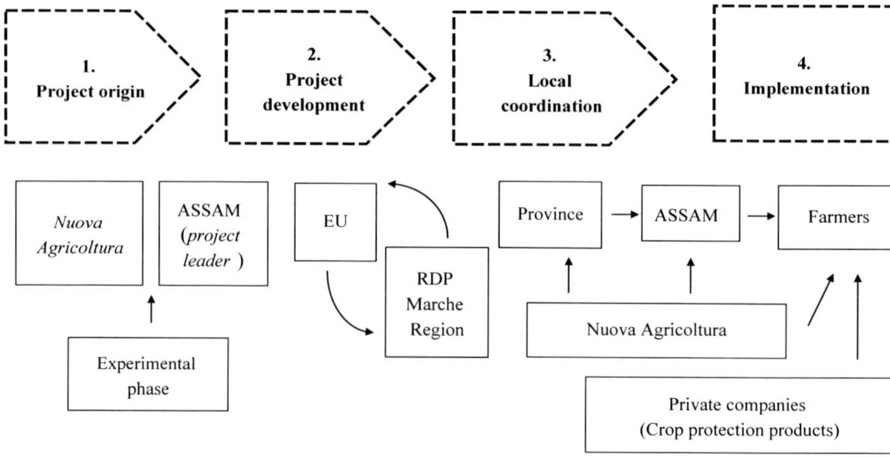

Fig. 5.2 Overview of the development of Valdaso AEA

According to the farmers, the project leader herself played a leading role in reinvigorating farmers' enthusiasm and in favoring their aggregation. Many farmers have also recognized an important role of leadership to two or three farmer 'pioneers', who according to those interviewed were equally important in developing this initiative. Moreover, although in Valdaso area the development of this collective strategy is mainly due to the Association *Nuova Agricoltura*, other local institutions and organizations have played a crucial role in the settlement and the implementation of the AEA.

The Regional and Provincial authorities, who were especially aware of the inefficacy of an approach focused on individual farms, were able to institutionalize and support the specific needs of local farmers related to agricultural practices and public goods protection. In other words, the bottom-up approach experienced through the activities of *Nuova Agricoltura* and of the project leader has been supported and coordinated by the regional and provincial authorities, in order to design and implement a mix of measures targeted to the local needs (Fig. 5.2). During the initial phase of the project *Nuova Agricoltura* developed the idea of adopting IPM techniques at a territorial scale in collaboration with ASSAM, in particular with the project leader, who was one of the key drivers of the development of this collective agri-environmental action.

In the second phase the support provided by the Regional administration was crucial, by setting the environmental goals to be achieved and by proposing the specific package of measures to be integrated into the Regional Development Programme. In the regional RDP the possibility of concluding AEA was already foreseen but there were no farmers interested in that kind of agreement. So, as the provincial authority of Ascoli Piceno came up with the request for an AEA, the regional administration has managed to embed this request in a viable agreement, by setting operational rules in the RDP, after the approval of the EU.

As highlighted by an official of the Marche regional government, the request was made explicitly by a group of local farmers associated to *Nuova Agricoltura*:

> We had a request from a group of local farmers who wanted to experience this innovative technique. ASSAM played an important role of technical assistance and contributed to disseminate the good results of the experimenting phase. When many farmers were convinced to act collectively, the Regional government, through the provincial authority, was quite happy to help them and to find a specific support for this action (INTERVIEW n.18).

Then, in the coordination phase (phase nr. 3), the Provincial authorities involved (initially Ascoli Piceno provincial authority and in a second phase also Fermo provincial authority) have been very important actors in building the local partnership of farmers and institutions and, above all, in translating the general guidelines of the regional RDP into local strategies as well as in coordinating the AEA at territorial level.

As described by those responsible for the Ascoli Piceno provincial authority, the coordinated strategy of *Nuova Agricoltura* and the local provincial authorities resulted in an inclusive decision-making system, whereby the farmers could also participate in the definition of the technical requirements of the measures to be applied:

> The project was born as result of a strong determination of local farmers, who have asked the provincial authority to act as intermediary with the Marche regional government to add value to their collective action. From them and from their strong desire was born the whole process, as a provincial government we acted only as a bridge with the regional government. (INTERVIEW n. 15).

The provincial authority of Ascoli Piceno was involved in the project by the farmers' association and by the project leader, since this institutional level was considered the most appropriate in dealing with the local needs. Local stakeholders agree that the provincial authority represents the most suitable institution to translate these needs into an agreement funded by the regional RDP. As will be further discussed later, respondents argued that the problems that emerged during the implementation of the AEA was due to the lack of formalization of the role of the Provincial authorities, since these institutions in the Marche region are not in charge of the agricultural policies, but they have been involved specifically for implementing this agreement.

Finally, during the implementation of the agreement at farm level (phase nr. 4 in the Fig. 5.2), ASSAM played an important role, by giving advice about IPM techniques and for the development and transmission of new knowledge among farmers. The project leader emphasized the role of the local network, which was crucial in the development of a collective project, and she defined the agreement as the result of a co-creation process:

> The agreement is the result of a co-creation process, where each actor involved has built something. It was a synergy of forces, a group of work (INTERVIEW n. 14).

5.2.2 The Co-management of the AEA

The territorial approach experimented in Valdaso is based on a broad involvement of local actors, not only during the development of the agreement, but also in managing the technical and administrative tasks of the project.

Table 5.1 summarizes all the actors that, at different levels and for different tasks, have been involved in the design, implementation and management of Valdaso AEA.

Table 5.1 The actors involved in Valdaso AEA and their role

EU and national level	
European commission	Approval and design of Rural Development Programme for Marche region
Regional government	Introduction of the AEA into the regional RDP
District level	
Ascoli Provincial Administration	Promoting and managing authority of the AEA
Fermo Provincial Administration	Managing authority (with Ascoli provincial authority) since 2010
ASSAM	Technical assistance, chemical analysis, information system
Crop protection products companies	Technical assistance
Farmers' organizations	Administrative assistance
Local level	
Nuova agricoltura	Development of the initiative, involvement of local farmers
Project Leader	Development of the local network of farmers, bridge with the local institutions
Local farmers	Experimentation of techniques and adoption at territorial scale

From the point of view of policy design, a central role has been played by the regional government, which has tried to set the regulatory framework for the territorial agreements in the regional RDP. It must also be observed that the Marche regional government in the past had already experienced a number of pilot environmental projects with a territorial/collective approach, including projects designed to reduce soil erosion. For several reasons these pilot projects have not been translated into effectively implemented operational projects, but they have been a strong basis on which to build the regulatory framework for the following territorial and area-based agreements. As highlighted by those responsible for the AEA at the Marche region, this strong methodological basis represented an important pre-requisite for translating the requests coming from the Association *Nuova Agricoltura* and from the provincial authority of Ascoli Piceno.

With regard to the local implementation of the AEA, the provincial authority of Ascoli Piceno has been identified as the most appropriate local authority for several reasons. Firstly, as argued by the founders of the agreement, the provincial authority is able to act as a bridge between the local dimension (farmers) and the regional dimension, namely between the design of the policies and the beneficiaries of the payments:

> A collective approach for the agri-environment must involve direct contact with the territory and with the people involved. The regional administration is too far from the local dynamics, does not have the right proximity to the needs of the territory. The innovation of this agreement is related to the territorial governance, since the provincial authority was able to promote a process of aggregation at territorial level. The municipalities cannot do it, since they do not have a district perspective of the actions needed; they are mainly focused on isolated and very local examples (INTERVIEW n. 17)

At the same time, a few problems have been highlighted related to the local management of the project.

Firstly, there is a problem relating to the lack of devolution in the managing of the RDP in Marche Region. Indeed, in some regions such as Tuscany and Emilia Romagna, provincial administrations are in charge of implementing the Local Rural Development Plan but it is not the case for Marche Region, and this had some important negative effects. As stated by an official of provincial authority of Ascoli Piceno:

> The AEA is managed by the provincial authority but we are not in charge of implementing agricultural policies, so we carried out several tasks for this agreement but officially we do not have the authority to do it. It would have been better if we could be involved to a large extent into the agreement, by following it step by step and to be a bit more present with the farmers, who usually call us to solve problems that we cannot face because the RDP is fully managed at regional level (INTERVIEW n.15).

As result of this, the provincial officers have emphasized that the activities carried out by the provincial administration does not have the financial resources for what concern the organization and management of the AEA, and some farmers have expressed some discontent of not being adequately trained and monitored by the institutions in the ways prescribed by the agreement.

Secondly, a barrier that was highlighted by the local stakeholders was the recent division of the territory covered by the agreement in two provinces: Ascoli Piceno and Fermo provinces. Indeed the Fermo provincial authority was established in 2004 but started its activities in 2009 (the involvement in the agreement started in 2010). Currently the river Aso is the natural boundary between the two provinces and this subdivision has created some problems in the management of the AEA.

Amongst the other actors involved in the management of the AEA, farmers and representatives of local authorities have also underlined that there has been an important support from professional organizations (Cia, Copagri, Cupa and Coldiretti), especially concerning the administrative support (farmers' files and applications):

> The province set up a so-called 'green table', where, together with *Nuova Agricoltura*, the official farmers' organizations were also involved. These organizations were important especially in providing helpful data on the farmers involved and in helping farmers to make the applications for the agreement (INTERVIEW n. 17).

Nevertheless, some respondents have emphasized that, while organizations have supported the project when it was already in place, they were unable to give it a push during the early phases, where they did not support adequately the initiative of *Nuova Agricoltura*.

The co-creation of the agreement also saw the active participation of ASSAM, which played a central role for what concerned the technical assistance the information system. Moreover, as it will be further discussed, the ASSAM center of Jesi carried out the analysis of the residues on the products in order to evaluate the effects of the AEA in terms of food safety.

In some cases the technical assistance provided by ASSAM was integrated with the advice provided by some companies specialized in crop protection products (Basf and Bayer) which were increasingly interested in commercializing products for IPM and especially for mating disruption. This interest is also due to the EU directive on the sustainable use of pesticides, that is setting increasing restrictions regarding the use of conventional products. This interest led to an involvement of these companies in the communication and training activities foreseen in the context of the AEA, which were also an important occasion for them to show their products that can be used for IPM.

This involvement has been reported after the implementation of the agreement, while in an initial phase the attitude of these companies was different, as described by a technician working for ASSAM:

> At the beginning these companies were not interested at all, since they had the convenience that farmers continued to use the conventional methods, because they were used to buying a lot of chemical products. The development of this agreement, together with the EU Directive, has been the main driver that changed their approach and they have been increasingly involved in spreading the IPM techniques amongst local farmers (INTERVIEW n. 20).

With regard to the local dimension of the agreement, namely the interaction of the institutional and technical actors with the single farmers, all the interviewed have underlined the crucial role that has been played by the ASSAM project leader:

> She has a strong capacity aggregation and she was crucial in enhancing the cohesion and cooperation amongst all the people involved in the agreement and she represented a key factor in convincing the most skeptical farmers to participate in this collective action. She has a direct relationship with everybody, she is a very capable and professional person and, above all, everybody trusted her (INTERVIEW n. 15).

This was confirmed by the project leader herself:

> I tried to convince everybody since at the beginning and I was involved in first person. I tried to change the farmers' attitudes to innovation as much as possible, and I made a strong effort to convince them to try the mating disruption, at least for 1 year; fortunately the first year the technique worked well for almost everybody, thus they were convinced to keep going with this project (INTERVIEW n. 14).

As already discussed, the project leader was also crucial for the administrative aspects, representing the contact between local farmers and institutions and she was definitely the most important actor that made the co-management of the AEA work.

5.2.3 *Communication and Marketing Strategies*

Several communication strategies have been associated with the AEA in order to increase the visibility of the agreement, which at national level is already acknowledged as one of the most innovative approaches to agri-environmental measures[1]. Moreover, farmers participating in the agreement have also tried to inform local communities on this initiative during the local fairs and markets where they commercialize their products, also through leaflets and brochures. According to the local

[1] Amongst the various initiatives that have been undertaken at the local level there are: a video on the IPM techniques and the agreement that which was financed by the provincial authority in the framework of a project on food education for the school, a few articles on the regional newspapers and an article on the magazine of the European Network for Rural development.

Fig. 5.3 The signs indicating farmers' adhesion to the AEA

stakeholders interviewed, one of the most important initiatives, which allows farmers to communicate to local consumers the role of the AEA and the results of IPM is the yearly agricultural show, which takes place every December in Ascoli Piceno (the show is denominated '*Gioielli della terra picena*'—Jewels of the Picena area). In the context of this event, together with the market of local products, several conferences and seminars are organized on several topics related to the agriculture and rural development of provinces affected by the Agreement (Fermo and Ascoli Piceno provinces). Some visibility was also offered by several professional organizations, for example, through the initiative *Campagna Amica* of Coldiretti.

In spite of this broad range of communication strategies, farmers and the other stakeholders involved in the agreement emphasized that during the initial phase one of the most important barriers of the initiative was the lack of a coherent and comprehensive strategy, since farmers were not able to differentiate the fruits on the market, which was as commercialized as the conventional ones.

During this phase the marketing and information strategies were mainly based on informal initiatives and non-coordinated informal initiatives. For example in several local stores it was possible to find signs with the writing: 'peaches produced with mating disruption' and, similarly, farmers installed *ad hoc* signs in their farms (especially farmers who do the direct selling on farm) in order to show that they adhere in Valdaso AEA (Fig. 5.3).

After this initial phase a coordinated action between *Nuova Agricoltura*, the regional government and local supermarkets aimed at overcoming this barrier and the AEA became part the regional strategy for food labelling: the '*QM-Qualità garantita dalle Marche*' (Guaranteed quality of Marche Region).

This was also due to the increasing appetite for quality products expressed by local supermarkets, especially for local fruits with low residues. This interest met the need expressed by the local farmers to enhance and valorize the innovation of the agreement in the supply chains, possibly through a collective brand (Fig. 5.4).

According to the farmers, the adoption of the QM label enabled them to attract more attention from consumers and at the same time to ensure greater food security, in light of the principles adopted by the European Commission.

Finally, the adoption of the QM label is a very interesting example of the economic and social effects that may result from a coordinated collective action which

Fig. 5.4 The QM label

was developed mainly for environmental reasons. Through the label local farmers have the opportunity to obtain higher prices on the market but at the same time have the opportunity to consumers their collective commitment to sustainability and to increase the reputation and trust of local communities towards their production practices.

5.3 The Involvement of Local Farmers

5.3.1 Farmers' Motivations and Attitudes

Among the factors that may strongly influence the adoption of an innovation in agriculture, the impact of innovation on production risks is very important (Abadi Ghadim et al. 2005; Marra et al. 2003). According to the farmers and technicians interviewed, the payment received by farmers involved in the AEA (about 650 € /ha) was an important incentive for farmers to face this risk and by testing the innovative farming practices by *Nuova Agricoltura.* As stated by two representatives of the provincial authority of Ascoli Piceno:

> I would say that, especially the first year, almost 40 % of the participants decided to adhere to the agreement because of the payments ... then it is essential to train people and teach them that the payment is something that they receive to change their farming practices, but also that the objective of policies is also spreading the adoption of convenient and profitable practices (INTERVIEW n. 15).
>
> Farmers are skeptical about new techniques, absolutely, but farmers may be persuaded by payments. In some way it is sad, but it works like that. Higher payments may convince them, but it is necessary to manage well such payments and above all, to put them in the right place, associated to the right practice that farmers are ready to adopt (INTERVIEW n. 17).

Thus, with regard to the involvement of farmers, the results of the research confirm what several economists have already demonstrated: the participation into collective

Table 5.2 Farmers' motivation to participate into the collective action in Valdaso

Individual/economic reasons	Social/environmental reasons
CAP payments	Reducing water and air pollution
Effectiveness of integrated agriculture methods	Increasing the tourist attractiveness of the area
Cost reduction (labour, crop protection products, fuel)	Positive effects of being involved in a local network of farmers (increasing learning opportunities and lobbying power)
Positive health effects	

action is strongly motivated by private benefits (Ayer 1997). Indeed, those interviewees not belonging to the Association *Nuova Agricoltura* argued that they participated in this collective action because they believed that they would gain from participation, both by obtaining the payments of the AEA and by saving in costs, due to the reduced use of chemical products and machineries.

Indeed, many farmers highlighted the issue that a reduction of the treatments with pesticides in the orchards as results of the IPM was also good because this contributed to reducing the production costs, since a smaller quantity of crop protection products was purchased. At the same time, the reduction in the use of crop protection products also resulted in a decrease in the use of tractors and other machineries as well as of the labour.

In addition, as already observed, another driver which influenced farmers in adhering to the AEA is the recent EU legislative framework, in particular the Directive on Sustainable Use of Pesticides 2009/128/EC, which aims at reducing the risks and impacts on human health and the environment related to the use of pesticides, by reducing the number of permitted chemical products for crop protection and by promoting the use of alternative pest management methods.

The most reluctant farmers were persuaded by the most motivated farmers of *Nuova Agricoltura* but, as they have highlighted, the efficacy of the new farming practices adopted played a significant role. In many cases local farmers were already looking for alternative methods of cultivation, since they had observed increasing resistance of pathogens to the conventional crop protection products, resulting in higher crop losses.

The bulk of farmers involved in the agreement declared that they were also motivated by reasons which go well beyond economic incentives, such as the willingness to reduce their health risks and to reduce the environmental impacts of their farming practices (Table 5.2).

Thus, the data collected through the interviews shows that the birth of the AEA is also the result of a growing environmentally awareness of local farmers, who increased their perception that more sustainable farming practices may play a key role for a transition towards a more sustainable development to the whole of the valley:

> We have understood that we may contribute to save the valley. It is clear that farmers can make a lot for the local environment. There are an increasing number of tourists here, they want to stay in agri-tourisms and they want a beautiful landscape and a safe and unpolluted

> environment. Houses have recently been renovated, new campsites have been opened, and farmers must contribute to this transition, by producing in a different way, and all together (INTERVIEW n. 23).

In addition, the majority of farmers also emphasized that the issue of health is crucial and that they are also adopting more sustainable techniques because they spend a lot of time within the orchards, and they want to work in a safer and healthier environment.

According to them, the most critical aspects are related to the fact that the IPM techniques, and especially the mating disruption, are knowledge-intensive techniques. This factor was particularly important, especially during the initial phase, when producers who had always used conventional methods did not have the knowledge necessary to manage an orchard which is not always 'protected' by chemical products.

Farmers have highlighted that these methods require additional organization and monitoring, both regarding the presence of insects in the orchards and regarding the weather conditions.

Box 5.1–Transect walks with the local farmers

Transect walks were carried out with five farmers involved in the AEA. The objective was to have a direct (and joint) field observation and exchange, in order to stimulate additional reflections and considerations on the local farming system and on the impact of the AEA. During the transect walks farmers extensively discussed their production activities, the main features of their farms (size, location, type of production, irrigation systems etc.) and, above all, the main problems regarding pest management and plant protection.

Compared to the transect walks of the 'Custody of the Territory' project (Chap. 4), where general observation of an entire area was done, in this case study transects were aimed at observing practices at farm level, and they were more focused on the characteristics of single farms. The discussion covered several environmental aspects of farming, such as soil erosion as well as soil and water management. Nevertheless, the most detailed descriptions were related to crop protection and more specifically to IPM. The discussions carried out during the transects allowed a deeper understanding of the main technical problems regarding the adoption of this technique, especially regarding its effectiveness for different species and cultivar, in different locations and according to different weather conditions. Farmers demonstrated an impressive knowledge of entomology and plant physiology and a clear perception of the main risks and opportunities derived from the adoption of the new practice. They also showed their enthusiasm and pride in farming in a healthier and safer environment and above all, of being pro-active members of a successful collective action.

5.3.2 Learning Opportunities for Farmers

In the case of Valdaso AEA, information exchanges and learning processes were central to the success of this collective action to many extents: from the adoption of the IPM techniques, to the dissemination of such techniques to other farmers.

Indeed, one of the most interesting aspects of the agreement is related to training and dissemination to farmers, since the measure 111 of the Regional RDP was used to finance on-farm visits as well as courses on IPM and administrative and procedural issues regarding the agreement, with the involvement of institutional representatives, ASSAM advisors as well as some representatives of companies specialized in crop protection products.

It should be stressed that the adoption of innovations by farmers is a dynamic learning process influenced by a broad range of issues, involving personal, cultural and economic characteristics as well as the characteristics of innovation itself (Pannell et al. 2006).

In the case of Valdaso AEA, the joint role of *Nuova Agricoltura* and ASSAM was crucial in creating a lively learning environment for farmers both through institutional and informal channels.

Firstly, a successful involvement of the local farmers was possible because of the informal meetings organized by *Nuova Agricoltura*. Furthermore, through the technical assistance of ASSAM, local farmers for the first time have been aware of the possibilities to change their farming practices towards more sustainable techniques. These meetings also represented significant occasions in which to discuss technical problems arising from the use of the advanced integrated techniques. Although at the beginning there was some skepticism about mating disruption, the positive results obtained by the farmers associated with *Nuova Agricoltura* resulted in a sort of domino effect, with the involvement of an increasing number of local farmers in the AEA.

IPM practices have then become part of farmers' knowledge, while dissemination of information by the ASSAM remained fundamental. The main instrument for dissemination of information is the weekly 'bulletin' sent to farms (by emails or fax), which can be also found in the noticeboards of the municipalities involved in the AEA. This bulletin is divided into agro-climatic zones and indicates to farmers very detailed weather forecasts with some information on the need for making chemical treatments or not, which problems may arise and how they can overcome them. The bulletin is also a means for informing farmers about additional possibilities (i.e. rural grants and funding) and other initiatives such as meetings with technicians of the ASSAM or other local events (fairs, local markets etc.).

Many farmers involved in the AEA recognized the bulletin as a very important tool since they usually follow strictly the indications provided by ASSAM:

> Farmers are in close contact with us, they usually wait for the bulletin to decide what to do, the bulletin gives them the direction, but if they think that it is necessary to make some treatments they ask us directly. We know the results of the traps both in farms that use mating disruption and in farms that don't use it, so we have a clear picture of the need of the area in terms of crop protection. For these reasons farmers strictly follow our advice (INTERVIEW n. 20).

Besides, ASSAM technical staff regularly visit farms involved in the AEA and monitor them to obtain information about the eventual problems they encounter with the IPM.

In some ways, through this agreement, ASSAM changed the traditional top-down approach of the technical advisory system of the area, traditionally based on private consultants working for chemical and pest control companies.

The role of a public advisory agency such as ASSAM was particularly relevant, since the technicians provided the assistance directly to farmers with the main objective of increasing the spreading of IPM at territorial level. These kind of intermediary actors are very important because in many cases they allows for the building of bridges between two different types of actors: the local institutions and the farmers involved in the project (Schneider et al. 2009). Indeed, amongst the actors involved, ASSAM technicians were the most aware that, in order for the collective action to be successful, it was necessary for certain types of knowledge to be created and communicated within the farming community.

In addition to this, one of the characteristics of this agreement was facilitating the exchange within the farming community and this was emphasized by the local stakeholders as one the key drivers of its success. Local stakeholders emphasized that the official meetings foreseen in the context of the agreement in many cases have been an opportunity for local farmers to meet and share point of view and experiences on the new techniques adopted. As stated by this farmer:

> The agreement has also been a good opportunity to know new people and farmers. During the meeting organized by the provincial authority we had the possibility of meeting several people who I did know before and that I met because I have joined the agreement. This has

> been very good for me, because I had the opportunity to increase the interactions and the exchanges with other farmers of my area. Thus, we can confront each other about the main problems related to the new techniques and the new farming practices (INTERVIEW n. 28).

A very important role was played by the pioneer farmers involved in *Nuova Agricoltura*, especially two of them who have a specific education in agronomy have tried to facilitate the exchanges between the most motivated farmers with the most reluctant ones.

Finally, the ASSAM agro-chemical centre of Jesi carried out the analysis of some samples of products in order to make public the differences, in terms of residues, of chemical products in fruits cultivated with both conventional and IPM techniques.

This centre has carried out the chemical analysis of the fruits produced by farmers both participating and not participating in the agreement. The selection of farms was based on the geographical location (through GIS) and later the ASSAM technicians contacted farmers to collect the sample in their orchards. The results demonstrated that the fruits produced by farmers adhering to the AEA had much lower residues of pesticides compared to the fruits produced in farms that did not participate into this agreement.

More specifically, ASSAM has analyzed samples from 37 farms, 24 of them using IPM and 13 not using IPM. In both cases the residues found in the samples were below the maximum allowed by law, but the farmers not using IPM had a higher percentage of samples with trace of residues of pesticides (78 %) compared to farms adopting integrated agriculture (57 %). In addition, the products from conventional farms had a higher percentage of samples with simultaneously multiple residues (21 % compared to 7 % for farms using IPM).

These results were presented in an open meeting, which was very successful because it showed to the farmers the good results obtained with the IPM techniques, making them aware of the substantial results of their commitment.

This confirms what was recently highlighted by the European Court of Auditors (2011), namely that the feedback on the impacts of agri-environmental schemes may improve farmers' awareness and understanding of the related environmental effects, especially where farmers have completely changed their practices. Indeed, ASSAM representatives underlined that farmers participated very actively in this event because they were looking for feedback on the efficacy of the technique they had adopted and expressed their willingness to make future analysis of their own products:

> We did not expect all those presences of farmers and, above all, the farmers who participated at the meeting were well motivated and informed, and this made us very happy. I think because they could see the results of the analysis, and then being aware that IPM is objectively important to harvest healthier products (INTERVIEW n. 19).

The important role of this initiative in motivating farmers, and in spreading the information on the effectiveness of the new techniques, was also confirmed by several farmers:

> It was very important to see that here have been some concrete results. Because the ultimate goal of the agreement is having healthier products and the results of ASSAM show that this

is the case. In the fruits that I have harvested last year there were no residues at all; in my neighbor's fruits the residues of pesticides were ten times below the conventional fruits. This is very encouraging . . . (INTERVIEW n. 24).

Thus, the communication to farmers of the reduction of the residues of pesticides in their products as a result of the IPM techniques was particularly important in order to increase the motivation of farmers in continuing to adopt the related practices. This evidence confirms what has been shown in several northern European countries, where it was observed that the understating of the benefits of agri-environmental measures usually results in a commitment of farmers to undertake the related management and, above all, it is more likely that the outcomes are to be successful and sustained in the long term (see Poláková et al. 2011; Herzon and Mikk 2007).

Finally, the case of Valdaso shows the success of an integrated strategy, since the provision of advice associated with the agri-environmental strategy was identified as one of the key factors of success. Indeed, the training activities, alongside the advisory system in place related to the adoption of IPM techniques, were particularly effective in increasing the awareness of the local farmers to the environmental priority of Valdaso areas as well as in increasing their skills in adopting more sustainable farming practices at a territorial scale.

5.4 Benefits, Barriers and Potentialities of the Agreement

5.4.1 Factors Affecting the Collective Action

The territorial agreement in Valdaso described above represents a quite innovative strategy, at least at the national level, where the collective dimension of public goods protection has been addressed at territorial level, by implementing innovative policy tools and institutional arrangements which favoured the support and the institutionalization of a grass roots collective action undertaken by a group of local farmers.

Table 5.3 summarizes the key factors which have determined the success of this collective according to the four categories described in Sec. 2.1: (1) the characteristics of the resource system, (2) the characteristics of the group of local stakeholders involved, (3) the local institutional arrangements and (3) the external forces (in terms of policies and authorities).

The increased awareness of local farmers regarding the severe air and water pollution of the area was identified as one of the main drivers of this collective action. This is particularly interesting also because the farmers who promoted the collective action are highly specialized in fruit production, and in the past they admitted to a very high use of fertilizers and pesticides.

At the same time a deeper analysis of the collective action allowed us to identify the internal issues of the initiative, such as the important role of a group of highly motivated farmers (associated in *Nuova Agricoltura*) and the crucial role of

Table 5.3 Factors affecting collective action in Valdaso AEA

1) *Resource system characteristics*	2) *Group characteristics*
Severe water and air pollution	Local leadership (ASSAM technician)
Intensive fruit production	Small group of motivated farmers (*Nuova Agricoltura*)
High specialization of farms	
High use of pesticides	Social capital (traditions of cooperation)
3) *Institutional arrangement*	4) *External environment*
Co-management of the AEA	Pre-existence of a broader territorial project (Valdaso Project)
Pre-existence of a local farmers association	
Involvement of a public extension service (ASSAM)	AEAs foreseen in the regional RDP
	EU Directive on sustainable use of pesticides

the project leader, who ensured the required bridge between farmers and local institutions, by setting both the technical and administrative requirements for collective agri-environmental schemes. As observed by Ingram (2008), the role of individual facilitators may be crucial in empowering, enabling, re-skilling and re-orienting farmers, in order to help them to think through what they want and how to achieve it.

In many cases local authorities' representatives have a clear understanding of the objectives of the schemes and how they should be implemented, but they may have a more limited knowledge on how the actions proposed may fit in with the practices adopted by local farmers. In the Valdaso area the role of the project leader was twofold. On the one side she was able to help local institutions to understand the needs of local farmers and to increase the institutional capacity; on the other side she coordinated the collective to carry out the administrative tasks required by the local authorities.

With regard to the main institutional arrangements that allowed the supporting and institutionalizing of this collective action, it is worth emphasizing the effective co-management of the local different actors, as described in Sec. 2.3, as well as the presence of a tradition of cooperation in the area.

Moreover, both farmers and technicians have emphasized that the success of collective action implemented to protect agri-environmental public goods is strongly related to the presence of a public advisory service. Indeed, in the case of Valdaso AEA one of the aspects most emphasized by the interviewees was the fact that the technical assistance provided by ASSAM was of a 'public nature' and 'disinterested' towards the selling of crop protection products. This is particularly important since in this area the technical assistance on the farm has been traditionally provided by advisors working for private companies. Thus, the role of ASSAM was perceived as crucial to obtaining advice on sustainable agricultural practices coherent with the public goods objective pursued through the AEA:

> Farmers have perceived that our advice is for the good of the farm and of the environment, because we do not have any private interests related to their business. Private companies cannot carry out our role, because they cannot be in favor of reducing treatments, and reducing the purchasing of crop protection products (INTERVIEW n 20).
> The interest in "public" goals can only be ensured by a "public" institution such as ASSAM. We can make advice for farmers but without any private interest, for example in selling products. I think that the involvement of our institutions was crucial for the success of this collective action (INTERVIEW n. 21).

The presence of this local (public) advisory system, which provided not only technical guidelines but also learning opportunities for famers, was very important in facilitating the sharing of information within the farming community.

With regard to the external factors, the presence of a broader project denominated 'Valdaso project' was also important, but the most relevant factors are related to the effective coordination mechanisms at institutional level. These mechanisms ensured an effective institutional support to this action bottom-up initiative, also by including a broader network of local actors involved.

Finally, innovative policy tools were implemented, tailored to the specificities of the area but also taking into account the changing legislative and policy scenario (i.e. the EU directive on sustainable use of pesticides). Also thanks to these external inputs, integrated packages of measures included in the RDP were specifically designed with the objective of addressing in a more coherent and coordinated way the environmental, social and economic issues emerging in the area.

5.4.2 Benefits and Barriers of Collective Action

According to the local stakeholders interviewed, compared to the traditional policy tools, such as the cross compliance and the traditional single agri-environmental measures, the collective action of the Valdaso area resulted in several positive effects. The observed benefits are of different types, both environmental and socio-economic. All these benefits are in some ways related to the local governance and on the institutional arrangements experienced. Indeed, the joint role of private and public stakeholders, together with the integration of different RDP measures in a territorial agreement, favoured the implementation of a coherent strategy more finely-tuned to local needs.

In Valdaso the protection of agri-environmental public goods was pursued through the direct involvement of farmers, by translating their willingness to increase the sustainability of their farming practices into an operational project with specific environmental targets to be reached at a territorial scale.

This approach was particularly innovative because, as highlighted by the ASSAM project leader, it was possible to reach not only environmental benefits, but also a transfer of knowledge to the territory:

> The main benefits of this project are related to its territorial approach, where the action is not addressed to individual farmers but to an entire area which can benefit both environmentally that in terms of knowledge; can you imagine the amount of knowledge that has been transferred to this area? A wide group of farmers in few years has moved to a new technique... I think it is great (INTERVIEW n. 14).
>
> I think that this project will go ahead in the future even without the payments, even if we have to carry out it at our own expense, because I think that people understood that it is possible to obtain important results by cooperating and collaborating (INTERVIEW n. 25).

Thus, an intangible but important public good produced by the collective action is the new knowledge which has been created and disseminated among farmers

Table 5.4 Benefits and barriers of collective action in Valdaso AEA

Benefits	Barriers
Right scale of public goods protection	Free riding
New farmers' attitudes	Control and monitoring
New farmers' approach to innovation	Unsatisfactory support from institutions (lack of communication and delays on payments)
Knowledge generation and learning	
Institutional capital and capacity-building	Higher transaction costs

participating in the agreement: farmers have emphasized that they have increased their knowledge to a greater extent as a result of this collective action (i.e. on the phenology of the insect, on agro-biodiversity and on plants cycles). It may be argued that the development of this new knowledge is essential and it is the foundation upon which to build a new model of policy intervention for the provision of public goods associated to agriculture. The longevity of this strategy is due to the fact that the new knowledge, which has been generated thanks to the rural development payments, will be accumulated into social and institutional capital that will also stay in place when the collective action would not be supported anymore by public policies.

Moreover, the set of motivations that have led farmers to join the agreement and the use of new techniques, along with the environmental and socio-economic results described above, have also resulted in a change of attitudes of farmers, which was perceived as one of the most important results, also from the point of view of policy development:

Another important benefit of the collective action in Valdaso is the fact that this initiative totally reversed the approach of farmers towards the innovation in agriculture. Indeed, the IPM techniques, and especially mating disruption, require a deeper knowledge of the agro-ecological systems. The collective action allowed farmers to increase their technical knowledge and this resulted in a more pro-active approach on the management of their activity, while in the past the adoption of innovation was too much driven by the private advisors of the companies specialized in crop protection products. As argued by a local farmer:

> Now I can bring my ideas, I am going to buy a chemical product only when I know what I need. Now I feel that I can decide for myself, the type of strategy that I wasn't to adopt for the crop protection in my farm. Before it was not like that, the sellers decided everything (INTERVIEW n. 25).

This new approach was spread also thanks to the contribution of ASSAM that, as described above, provided farmers with a public technical assistance and not oriented to the selling of crop protection products (Table 5.4).

With regard to the main barriers of the Valdaso AEA, the institutional stakeholders involved (representatives of the regional and provincial authorities) have also highlighted the problem of the high administrative costs in supporting and managing this territorial initiative.

As described in sec. 2.2, in many cases collective action, compared to individual action, may present higher transaction costs related to search costs, bargaining costs and monitoring and enforcement costs. Indeed, whilst the devolution of powers to local institutions is deemed to implement more targeted strategies, in many cases the complexity of the different and interconnected levels of governance involved in territorial agreements may result in an increase of costs related to the management and coordination of the collective actions.

In Table 5.5, these types of transaction costs are associated with the different actors involved. Even though this is not a quantification of the specific costs, the table shows the types of additional costs associated with this collective action and how they are distributed amongst the main actors involved.

As will be further discussed, while some of these costs are not currently financed through the RDP, a more effective support to collectives to ensure territorial agri-environmental strategies should receive additional funding for the initial capacity building process as well as funding for the coordination, management and group activities. This is true especially for farmers associations, but also for local institutions that are currently excluded from this type of funding, such as provincial authorities in the case of Valdaso AEA.

The barrier of higher transaction costs in Valdaso was overcome thanks to the strong commitment of farmers associated with *Nuova Agricoltura* and of the representatives of provincial authorities. Moreover, according to the respondents, some costs were easily addressed thanks to the relation of trust and reciprocity amongst the majority of farmers, by reducing the costs related to time and effort spent monitoring others, time spent at meetings, as well as the costs of gathering information. Similarly, according to the local stakeholders, the increased social capital deriving from this collective action resulted in a reduction of transaction costs for the local institutions, such as the costs of identifying relevant stakeholders, gathering information and enforcing sanctions.

With regard to the other barriers to this collective action, the technicians working for ASSAM emphasized that there were different levels of involvement amongst farmers, with some examples of opportunistic and free riding behaviors[2]. According to the representative of local institutions there have been several cases of farmers who adhered to the agreement only to receive the payments, but in reality continued to use chemical products for crop protection.

As a representative of the provincial government pointed out, in this regard the role of social control was important, since the farmers themselves, who know each other, indicated the possible free riders (who joined the AEA only to receive the payment) also because they were damaged by these opportunistic behaviors (as already observed the effectiveness of the mating disruption is highly dependent on the possibility of using this techniques on an unbroken piece of land).

This social control is the result of the relationships of trust and reciprocity which were created within the local farmers' community: many farmers highlighted that one of the most the positive effects of the AEA was the fact that they learnt to collaborate and cooperate to achieve common goals. Indeed, IPM requires stronger analytical

[2] As described in Sec. 2.2, another important barrier of collective action is the presence free riding behaviors, especially when a large group of participants is involved, such as in the case of Valdaso AEA.

Table 5.5 The transaction costs in Valdaso AEA

Transaction Costs	Examples	Nuova Agric.	Farmers	Project leader	ASSAM	Provincial authorities	Regional authority
Search Costs	Costs of identifying relevant stakeholders	✓		✓		✓	
	Costs of gathering information	✓		✓	✓	✓	✓
	Costs of identifying funding source for collective action			✓	✓		✓
Bargaining Costs	Time spent at meetings	✓	✓	✓	✓	✓	
	Effort expended in verbal and written communications			✓	✓	✓	✓
	Costs of acquiring support from external agencies			✓		✓	✓
Monitoring and Enforcement costs	Time and effort spent monitoring others			✓	✓		✓
	Employment of external monitoring				✓		✓
	Costs of enforcing sanctions						✓

skills and deeper understanding of agro-ecological principles as well as a higher cooperation between farmers, and these aspects may result in the creation of social and human capital (Pretty and Ward 2001).

Additional barriers related to this collective action are related to the process of institutionalization and to public support to the initiative and, more broadly, to the difficulties that could be encountered when trying to scale up or extend this initiative to other areas.

The founders of the agreement for the regional administration emphasized the problem of controls and monitoring, especially in relation to the environmental impacts of the agreement. Indeed, the ecological processes involved may result in positive environmental effects (reduction of air and water pollution) only in the medium-long term and it is very difficult to associate the provision of agri-environmental public goods to the results of the AEA.

The methodologies currently used for monitoring and evaluating the effects of RDP measures are largely inefficient and this is particularly true when more complex strategies are implemented, where several interlinked effects are provided, which in many cases are multidimensional (environmental effects as well as social and cultural effects).

With regard to the institutional support to the AEA, this agreement has certainly provided a good opportunity to increase the capacity building within the local institutions, since it was an opportunity to explore and experiment with a new approach in designing and delivering rural development policy measures. At the same time, in many cases the problems such as the division of the area involved into the agreement into two provinces and the lack of authority of provincial administration on agriculture, resulted in poor communication and valorization of the results, and some farmers participating in the agreement did not feel adequately supported and informed by the institutions.

In many cases the delay of payments was also perceived as a strong barrier, especially for less motivated farmers who decide to experiment with the new techniques, and the public support was crucial in the initial phase as insurance against the related risks.

According to the local stakeholders, it would be necessary to overcome all these limitations related to the institutional support to this initiative, in order to broaden and extend this approach and to develop such measures as a more coherent regulatory and institutional framework which could support more effectively this kind of agreement.

5.4.3 A Collective Approach to Agri-environmental Measures

Valdaso AEA was designed to protect soil and water from pesticides and nitrates pollution at river catchment level, through methods of production with low environmental impact. The project represents an innovative strategy to pursue multiple agri-environmental objectives, since an integrated suite of measures was implemented to increase the provision of agri-environmental public goods (water and soil quality, cleaner farming practices and healthier products) but also for reframing farmers' behaviors, attitudes and knowledge through a set of capacity-building initiatives and learning opportunities.

This case study shows that a bottom-up strategy developed by informal and local associations may have the potential of being formalized and institutionalized into the mainstream policy framework.

The CAP is highly concentrated on the approach of cross-compliance, which is considered the most cost-effective strategy in delivering environmental goods at large scale, especially in areas characterized by intensive and productive agricultural systems. On the opposite side, the AEA implemented in Valdaso shows the possibility of exploring bottom-up and collective approaches which, through innovative institutional arrangements and integrated policies, may protect public goods at landscape level and also in areas characterized by intensive agricultural production.

As shown by the case of Valdaso, intensive farming systems may also be reoriented to the provision of public goods through decentralized strategies based on innovative policy tools and institutional arrangements which may prove more effective in addressing local needs and environmental priorities. Unlike cross-compliance, which sets a series of compulsory requirements that must be followed in order to receive direct payments, a collective approach to agri-environmental action, such as the integrated delivery analyzed here, focuses more on a pro-active engagement of farmers in the provision of public goods. The case study also demonstrated that in the most productive areas, in many cases, farmers perceive the need for adopting innovative farming practices, in order to decrease the impacts of their activities on the environment.

The farmers' sense of belonging to their own community and the relationships of trust amongst local farmers were also developed as a response to IPM, since these techniques require a territorial scale to be applied. Moreover, as highlighted by Mantino (2011), through a territorial approach for public goods provision it is possible to provide not only environmental goods, but also a combination of economic and social effects, which would be difficult to achieve with a more traditional approach focused on land management practices adopted at farm level. Thus, in the Valdaso area a 'virtuous circle' between IPM and collective action was observed: the effectiveness of IPM was enhanced by the collective action but at the same time the adoption of IPM techniques encouraged farmers to aggregate themselves to get effective results. In this 'virtuous circle' both environmental and socio-economic benefits were produced (Fig. 5.5).

The main policy innovation of the AEA is related to the package of RDP measures involved, since the provincial authority of Ascoli Piceno acted as intermediary to push the regional government of setting eligible measures featured in the local strategy for promoting integrated agriculture. This package of measures was then transformed into a territorial agreement which could take into account the technical requirements for integrated agriculture but also the main requirements of the local farmers in terms of advisory, learning and networking.

The integration of RDP measures into a territorial agreement experienced in Valdaso was facilitated from some kind of 'devolution' to local institutions to incorporate

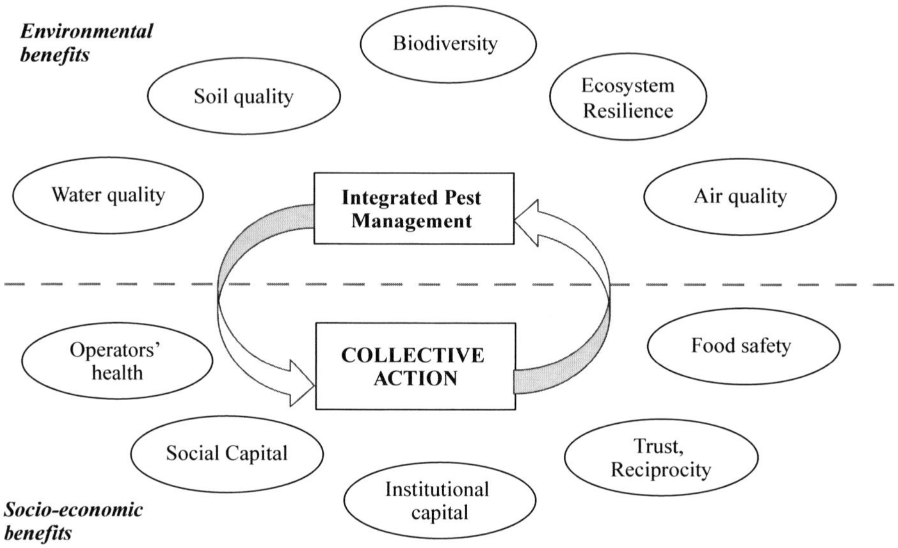

Fig. 5.5 The environmental and socio-economic benefits provided through the Valdaso AEA

integrated strategies into their territories. As suggested by the experiences in other countries (ENRD 2012), the creation of an attractive mix of measures (with training, technical assistance, compensations) in a project for farmers, is vital to promote participation. To summarize, in the context of rural development policies the success of collective strategies seems related to the development of specific institutional arrangements as well as to an integrated approach to delivery, where different RDP measures are used in combination to achieve broader and more effective results.

The experience of Valdaso, being the first of the 2007–2013 programming period, provided an opportunity for the region to test innovative administrative procedures, by facilitating the implementation of collective strategies for agri-environmental public goods in other territories, such as the Area Programme for biodiversity, which was launched in 2011[3].

At the same time, the local stakeholders have also highlighted several barriers related to the possibility of better integrate this collective approach into the policy framework of the CAP, which are mainly related to the local institutional arrangements and to the policy instruments currently in place. The experiences of the Integrated Territorial Projects in Italy, quite widespread during the 2000–2006 programming period, revealed several shortcomings, such as a lack of robust selection of participants, lack of integration amongst different operations, insufficient multi-sector approaches, and inadequate partnerships (Formez 2009). According to Mantino (2010, 2011), these limitations derive from two main factors:

[3] The Area Programmes for Biodiversity are mainly led by Natura 2000 payments with the support of additional RDP measures (125, 211, 214 and 216). These projects were developed by large consultation and participation of local farmers living in the protected area and Local Authorities.

- The resistance of the public administration to abandon the traditional method based on the single measures delivery. In fact single operations were selected within ITPs on the basis of separate and timely differentiated calls;
- A series of local coalitions did not act accordingly to an integrated plan, but simply followed the principle of capturing financial resources from public programmes. These can be considered only as instrumental coalitions.

The Valdaso AEA represents an interesting case where these limitations were, at least in part, successfully addressed. With regard to the first limitation, in Valdaso the public administration has tried to set innovative procedures which take into account farmers' needs and, above all, has also tried to incorporate the capacity building measures with agri-environmental measures for the procedural and administrative aspects. In addition, it may be argued that Valdaso AEA was the result of a grass roots collective action adopted for a more effective delivery of public goods and not the result of an instrumental coalition aimed at capturing financial resources.

From a policy development perspective, the main limitations highlighted by the representatives of the local authorities responsible for the implementation of the RDP in Marche region were mainly the lack of flexibility and autonomy in managing the RDP funds at the local level. This is particularly evident in relation to the integration of funding amongst the agri-environmental measures and the coordination and management activities necessary for territorial agreements (i.e. design, implementation and monitoring costs of the AEAs cannot be financed through the RDP, but must be financed by the local agency in charge of the coordinating the agreement). As observed by Mills et al. (2010), usually the development of collective strategies at territorial level requires additional funding for the initial capacity building process as well as funding for the coordination, management and group activities.

Finally, according to the local stakeholders, rural development policies usually lack flexibility to efficiently support spontaneous and endogenous initiatives, also because policies implemented for public goods in many cases are focused on administrative borders and are not tailored to specific territories. From this perspective, while in the Marche Region the RDP is directly implemented by the regional government, according to the local stakeholders, a sub-regional level implementation could have facilitated a more effective coordination at territorial scale, with higher environmental and socio-economic benefits.

References

Abadi Ghadim AK, Pannell DJ, Burton MP (2005) Risk, uncertainty and learning in adoption of a crop innovation. Agricultural Economics 33(1):1–9

Ascoli Piceno Province (2009) Accordo agroambientale d'area per la tutela delle acque e dei suoli da fitofarmaci e nitrati area Valdaso. Executive Project

Ayer HW (1997) Grass roots collective action: agricultural opportunities. J Agricultural Resour Economics 22(1):1–11

ENRD (2012) Coordination committee focus Group delivery of environmental services. Final report, December 2012, Brussels

European Court of Auditors (2011) Is agri-environment support well designed and managed? Special report no. 7. Office for Official Publications of the European Communities, Luxemburg
Formez (2009) Cooperazione fra attori e innovazione amministrativa: i primi risultati dei Progetti Integrati Territoriali nelle regioni Obiettivo 1. Dati e Tesi essenziali, Centro di Competenza "Politiche e Strumenti per lo Sviluppo Locale", Roma
Herzon I, Mikk M (2007) Farmers' perceptions of biodiversity and their willingness to enhance it through agri-environment schemes: a comparative study from Estonia and Finland. J Nat Conservation 15(1):10–25
Ingram J (2008) Agronomist–farmer knowledge encounters: an analysis of knowledge exchange in the context of best management practices in England. Agriculture Hum Values 25(3):405–418
Mantino F (2010) Understanding delivery mechanisms in EU rural development policies: an institutional approach. Paper presented at the 118th Seminar of the EAAE "Rural Development: governance, policy design and delivery". Ljubljana, Slovenia August 25–27
Mantino F (2011) Developing a Territorial Approach for the CAP. A discussion paper, Institute for European Environmental Policy, London
Marche Region (2007) Programma di Sviluppo Rurale 2007–2013. Marche Region
Marche Region (2008) Disposizioni attuative del Programma di Sviluppo Rurale 2007–2013 Reg. (CE) n. 1698/2005. Accordi agro-ambientali d'area per la tutela delle acque e dei suoli da fitofarmaci e nitrati. Marche Region, Ancona
Marra M, Pannell DJ, Abadi Ghadim A (2003) The economics of risk, uncertainty and learning in the adoption of new agricultural technologies: where are we on the learning curve? Agric Syst 75(2–3):215–234
Mills J, Gibbon D, Ingram J, Reed M, Short C, Dwyer J (2010) Collective action for effective environmental management and social learning in Wales. Proceedings of the 9th European IFSA Symposium, 4–7 July 2010. Vienna, Austria
Pannell DJ, Marshall GR, Barr N, Curtis A, Vanclay F, Wilkinson R (2006) Understanding and promoting adoption of conservation practices by rural landholders. Australian J Experimental Agriculture 46(11):1407–1424
Poláková J, Tucker G, Hart K, Dwyer J, Rayment M (2011) Addressing biodiversity and habitat preservation through Measures applied under the Common Agricultural Policy. Report Prepared for DG Agriculture and Rural Development, Contract No. 30-CE-0388497/00-44. Institute for European Environmental Policy, London
Pretty J, Ward H (2001) Social capital and the environment. World Develop 29(2):209–227
Schneider F, Fry P, Ledermann T, Rist S (2009) Social learning processes in Swiss soil protection—the 'From Farmer—To Farmer' Project'. Hum Ecol 37(4):475–489

Chapter 6
Collective Action for Public Goods: Institutional and Policy Innovation

Abstract Chapter six, by comparing the main findings of the two case studies analyzed in chap. 4 and 5, provides a more detailed theoretical discussion on the role of collective action in the provision of agri-environmental public goods. The institutional and policy implications of supporting and implementing collective action are discussed, also in the light of the current debate on the innovation needed to re-orient the EU Rural Development Policy towards a more effective provision of public goods in rural areas.

Keywords Institutional arrangements · Policy innovation · Rural development policy

6.1 The Determinants of Successful Collective Action

Amongst the main objectives of this book is the providing of some insights into the internal dynamics of collective action associate to agriculture, but also exploring the main determinants and drivers that allow collective action of effectively providing or protecting agri-environmental public goods at landscape scale.

The two case studies analyzed in Chap. 4 and 5 differ to large extent in terms of the type of collective action involved, since in the first case study (Custody of the Territory) the local action was mainly developed and coordinated by a local public agency (coordination), whilst in the second case study (Valdaso AEA) the collective action was directly led by farmers and later institutionalized and supported by the local institutions (cooperation). Moreover, there are also several differences between the two territorial contexts and, above all, between the agri-environmental issues addressed and the strategies implemented.

The main difference is related to the different institutional arrangements of the two initiatives but also to the different agri-environmental issues addressed. 'Custody of the Territory' is a project led by a local public body (Mountain Community Media Valle del Serchio) that has set incentives and supported a collective approach to the hydro-geological management of the district. This approach has resulted in a high dependence on the institutional role of the local authority and in its capacity of setting and managing efficacy of the PES with farmers. This institutional role, together with the increasing use of IT tools (i.e. IDRAMAP) has resulted in a more

DOI 10.1007/978-94-007-7457-5_6,

Table 6.1 The determinants of success of the collective action in the two case studies

Factors	Custody of the territory	Valdaso AEA
Social capital	Strong bonding social capital, weak bridging and weak linking social capital	Strong bonding, bridging and linking social capital
Leadership	Re-skilling farmers, intermediary person between local and regional institutions	Coordinating activities, creating learning opportunities for farmers
Right scale	Action implemented to provide environmental services at district level	Action implemented to protect environmental public goods at territorial level
Institutional arrangements	Co-production of knowledge and services	Coordination mechanisms and learning opportunities for the involved actors. Role of Nuova Agricoltura and project leader
Information and communication	2.0 web tools and meetings, participatory events with local communities	Emails, meetings, bulletin

technocratic approach to the delivery of environmental services, with great efforts towards implementing a structured and efficient information and communication system.

On the opposite, in the case of Valdaso, public institutions have been indirectly involved in a farmer-led collective action, by supporting a bottom-up approach in order to facilitate the adoption of integrated agriculture at territorial scale.

Nevertheless, it is also possible to recognize several common points between the two case studies, such as the strong focus on the knowledge and learning dimensions and the efforts of the local institutions in involving farmers in the decision making process. Table 6.1 synthesizes the key determinants for the two collective actions analyzed in the study.

As described in sect. 2.4, social capital is a crucial factor for successful grass roots initiatives. Indeed, social capital can help to overcome several problems associated with the implementation of collective action, especially in initiatives where a large and heterogeneous number of stakeholders is involved (Paavola and Adger 2005; Pretty 2003). The case studies analyzed in the book show that a collective approach may foster trust and reciprocity amongst the local stakeholders and how this, in some cases, may result in reducing transaction costs and increasingly the effectiveness of the agri-environmental strategies.

In particular, in Media Valle del Serchio (Tuscany), through the project Custody of the Territory, the local authority strengthened the bonding social capital amongst the farmers, technicians and the local community but, in some ways, failed to increase the bridging and linking social capital, and this resulted in a weaker scaling up of the project, which did not emerge as model of intervention in the surrounding areas and was not replicated in other areas.

On the opposite side, in the Valdaso area the association Nuova Agricoltura encouraged both bonding as well as bridging and linking social capital and this determined a broader transition to more sustainable agricultural practices at territorial level. This confirms what has been observed by Berkes (2009), namely that there is a potential for producers' organizations to play a role as bridging organizations, by facilitating open dialogue between producers at the farm level and policy makers at the government level. The case of Valdaso also highlights that with an appropriate capacity building programme a less formal and structured organization may provide platforms for information exchange and, above all may enable self-organization and collective action among local producers.

These local initiatives also show that the cultivation of trust and reciprocity amongst the participants of the collective action is usually the result of strong leadership. The key role of a leader was especially observed in the case of Valdaso AEA, where the project leader (the ASSAM technician) acted not only as an extensionist promoting new technologies but also as a key actor who was able to bring about the environmental re-skilling of farmers and, above all, acted as an intermediary between farmers and local institutions. This crucial role of the project leader enhanced farmers' confidence regarding the possibilities to undertake collective action for adopting sustainable practices at territorial level. Similarly, the role of the project coordinator in 'Custody of the Territory' was crucial in managing the advisory system of the project, as well as in coordinating the collective action and in enabling the technicians of the local authority to provide technical and administrative advices to local farmers for the delivery of environmental services.

Thus, these examples confirm the crucial role of leader personalities, who may represent the driving forces in forging alliances among public and private actors by initiating innovative projects to counteract the social and economic problems perceived as pressing by local population (Petrick and Gramzow 2012).

Another important key factor that determines the success of collective action is the issue of scale. Indeed, the effectiveness of the environmental action usually depends on the achievement of sufficient scale, such as the size of the areas interested and the continuity of the action across the territory. Since single landholders cannot satisfy these two conditions, in many cases a collective and a territorial approach is needed. The case studies analyzed here demonstrate that the success of collective action is due to an approach which goes beyond not only the boundaries of single farms, but also to the administrative boundaries of the territory where the collective action takes place. In order to provide public goods through collective action it is necessary to adopt a landscape approach, where the coordinated action is tailored to the natural resources to be managed and to the agri-environmental public goods to be provided, and not to the administrative boundaries and to the administrative roles of the different public bodies. It should be observed that in the two case studies this condition was not fully achieved. The main reasons are related to the lack of efficient coordination mechanisms amongst local institutions, to the high transaction costs that would have been incurred in increasing monitoring and enforcing at different scales, as well as the lack of flexibility of the policy tools used to support the collective actions.

In spite of these limitations, the case studies show that successful collective actions for public goods may be supported by innovative institutional arrangements. Indeed, as described in sect. 2.3, the success of collective action is usually based on co-management and co-production, whereas the allocation problem of public

Table 6.2 The information systems in place in the two case studies

	Custody of the territory	Valdaso AEA
Communication local institutions/farmers	Meetings and IT tools (IDRAMAP)	Meetings and IT tools (bulletins, newsletters)
Communication within the farmers communities	Meetings, workshops	Meetings, in-farm visits
Advisory system	Joint inspections farmers/technicians IDRAMAP	Specific technical assistance (ASSAM)
Involvement local communities	Participatory events	Newspapers articles, panels, QM label

goods is not totally left to the market or to the government, but mixed solution are experimented with.

From this perspective, the Custody of the Territory project shows that innovative PES schemes, based on co-production, may represent an example of mixed public-private arrangements which may deliver environmental services more efficiently. This approach is aimed at developing a pro-active role amongst farmers, by changing the logic of the PES and by developing a system based on co-production, which was perceived as a more effective way to deal to the flooding risks problems rather than systems based on centralized information systems and on pricing.

In the case of Valdaso AEA, on the other hand, the participation of a broad set of rural stakeholders determined a territorial strategy for public goods protection, based on shared responsibility and co-management amongst private and public actors.

The coordination and the cooperation between the actors involved in collective action are particularly important also with regard to generation, validation and exchange of information. As highlighted through the case studies, the success of collective action is usually related to an efficient information and communication system specifically addressed to increase the effectiveness of the collective action concerned. This system must make clear the required innovation, the new practices to be adopted as well as the distinction between private and public goals and the way such actions are coordinated and financed at territorial level (Table 6.2).

It may be argued that an efficient exchange of information and innovative communication system allows for the mobilization of the intelligence and the creativity of the rural stakeholders, by determining an increased provision of public goods. From this perspective new information technologies (IT tools such as 2.0 web sites, social networks, emails, etc.) may facilitate the information exchanges, new forms of education and training and, above all, increasing the transparency and the effectiveness of the strategies specifically implemented for increasing the provision of agri-environmental public goods at landscape level. Indeed, the case studies show that IT tools may provide farmers and citizens with clear information regarding the action undertaken.

Finally, as will be discussed in the next section, an important factor that determines the success of collective action is also related to the capacity of the involved actors of making better use of the knowledge and the experience of farmers for policy design and implementation.

6.2 A Pro-active Engagement of Farmers

The case studies described in the book allowed also for the exploration of the drivers that affect farmers' participation in collective action, by analyzing the main behavioral factors as well as the main implications in terms of technical and social innovation that must be generated in order to increase the participation of farmers into collective strategies for public goods provision.

The study shows that the delivering of agri-environmental public goods at territorial scale is a very complex challenge, which involves a relevant shift related to farmers' attitudes, motivations and norms, which must be translated into new farmers' behaviors.

Indeed, as observed in the case studies, financial incentives are not the only drivers to be considered when analyzing farmers' behavioral drivers, but motivations (psychological factors, such as attitudes) and norms also play a very relevant role.

The data collected in the field shows that in order to influence farmers' behaviors it is necessary to overcome the traditional linear 'top-down' transfer of technology. Indeed, more complex and 'landscape' problems, as those involved in delivering of environmental services or providing public goods, usually require integrated suites of practices to be adopted in an innovative and collective way.

A collective approach to agri-environmental issues requires a new pool of skills and knowledge that is usually the result of a wider spectrum of ideas and points of view, and usually it is necessary to integrate information from various sources. As shown in the case studies, the knowledge and the skills are the products of joint learning deriving from the interactions of several stakeholders who are directly involved in collective action, but who are also external to the projects.

The difficult challenge of spreading new technology and new knowledge in the case of collective action for agri-environmental public goods, is related to the extent to which a system is able to translate recommendations that derive from a 'landscape' environmental problem to suitable and attractive recommendations at farm scale level.

From this perspective, the case studies analyzed allowed us to explore how it would be necessary to reconfigure the approach to knowledge and innovation in order to implement successful collective action for public goods provision. The main elements for a pro-active engagement of farmers in collective action, which were identified through the case studies, are synthesized in Table 6.3.

The case studies show that one of the key determinants for a pro-active engagement of farmers into collective agri-environmental actions is the presence of an efficient public advisory system. Indeed, the technical experts of ASSAM involved in the collective action in Valdaso AEA were particularly relevant as intermediary persons who could build bridges between two different types of actors: the local institutions and the farmers involved in the projects. The advice provided by technicians of a public institution compared to those provided by private companies, resulted in an approach more oriented to the effective adoption of sustainable practices at territorial level compared to private advisory services, which usually provided profit-oriented

Table 6.3 The determinants of a pro-active engagement of farmers in collective action

	Custody of the territory	Valdaso AEA
An efficient public advisory system	Re-orientation of farmers' skills, joint learning farmers-technicians	Non-profit oriented advisors
		Bridges between local institutions and farmers
Valorization and re-building of farmers' local knowledge	Valorization of local knowledge regarding the environmental priorities, the location and conditions of the natural resources	Valorization of technical knowledge of farmers on the local habitats and on the techniques more suitable for the local conditions
Learning opportunities	Co-production of knowledge, social learning on technical and procedural/administrative issues	Social learning and increased exchanges within the farmers' communities
Building new identities for farmers	Farmers as custodians: landscape stewards and early intervention networks	Landscape stewards, enhanced role in ensuring water and air quality
Involvement of farmers into the decision making process	Involvement of farmers in planning the environmental services, public intervention driven to the environmental priorities identified by farmers	Involvement of farmers in designing collective agri-environmental strategies (required practices and related costs)

advice to farmers (i.e. companies specialized in crop protection are mainly interested in selling products to farmers).

Similarly, with regard to the collective action in Tuscany the public profile of the local agency was determinant as an advisory system, which aimed at re-orienting farmers' skills towards the delivery of environmental services.

Moreover, both case studies show that an efficient advisory system based on social learning and farmer-to-farmer interactions may positively influence co-operative behavior, and may help to re-design institutions and policies based on collective action. As discussed in sect. 2.5, in order to increase the effectiveness of collective action for public goods provision through agriculture, the innovation system must be reformed accordingly, and social learning and co-production may represent important approaches to shift from private goals (such as productivity increase) to public goals (such as environmental care).

From this perspective, it has been observed that a stronger engagement of farmers into collective action is possible by valorizing and re-building farmers' local knowledge. The evidence collected in this study emphasizes the need for increasing farmers' awareness of the environmental issues in rural areas that in some cases means favoring the recognition by farmers of the relation between their farming practices and the related environmental outputs. To achieve this objective it is necessary to overcome the top-down approach of technological transfer from technicians to farmers, by increasing the participation of farmers on the basis of their local and context-specific knowledge.

The research also shows that the local knowledge of farmers in many cases should be the basis on which to build new knowledge, oriented to the provision of public goods and services. Thus, as argued by Vanclay and Lawrence (1995, p. 125): 'farmers' local knowledge to solve problems that are new to their experience, such as many environmental problems, is unlikely to be successful. The insidious nature of such problems means that farmers may still not recognize them—even after extensive damage might have occurred. While it is possible that many traditional problems may be solved with new extension methods, new problems, particularly environmental problems, may be best dealt with through a combination of new and traditional extension'. Indeed, while a traditional approach to innovation usually fails to address in a coherent way the collective dimension of emerging environmental needs, participatory and bottom-up strategies may also fail in balancing community and personal needs. As argued by Black (2000, p. 496), 'while participatory and group-based approaches to agricultural extension have various advantages when they are well implemented, they should not be regarded as the one and only strategy that can or should be used to facilitate the adoption of sustainable farming systems. Belief in a 'participation fix' may be just as naive as a belief in a 'technology fix'. Thus, with regard to the provision/protection of agri-environmental public goods, the main challenge for the Agriculture Innovation System is combining the new technologies with the traditional and local practices. The challenge is finding a solution that may be increasingly recognized by the farmers' communities and may have positive effects in the medium-long terms. Achieving this objective is thus necessary to create learning opportunities for farmers, which, as observed in the case studies, could be based on joint learning and co-production of knowledge.

As argued by Berkes (2009, p. 1695) when managing environmental resources through a collective approach, the different actors need to work and think together, and deliberate to generate new knowledge or make sense of knowledge from different sources. In the case studies analyzed in this book, the learning opportunities for farmers were not related only to the technical requirements and on the innovative (and more sustainable) farming practices. On the contrary, such learning opportunities were also designed to increase the knowledge in terms of administrative procedures, scientific data, information systems and policies. Moreover, the case studies analyzed here show that in many cases local technicians and advisors are most aware that for the collective action to be successful it is necessary that certain types of knowledge would be created and communicated within the farmers' communities. At the same time, this research also shows that the process of innovation is not limited to the farm level, but is open to the interactions of farmers with wider networks, where collective learning at territorial scale are crucial issues for an effective provision of public goods.

Indeed, the provision of agri-environmental goods usually involves complex mechanisms of delivery that may be learnt jointly through the interaction amongst public bodies, private actors, farmers and farmers' associations. This integration of different types of knowledge (farmers, experts and academic) is not only better for dissemination, but may also bring better solutions on the field. In terms of public goods and environmental services provided by farmers, social learning and co-production of knowledge may be necessary processes to overcome the contradictions between sectionalized forms of knowledge, especially between the private/public goods provision through farming activities.

Finally, the adoption of innovative practices by farmers as well as their involvement in delivering public services (e.g. landscape management practice), are strongly related to the production of new knowledge and the spreading of innovation. As argued by Deuffic and Candau (2006, p. 565), the achievement of new environmental objectives will involve significant changes in farmer's practices: a technical change but also a more structural shift regarding their professional identity and their role in the management of the territory. The present research confirms this statement and it underlines that an innovative perception of farmers on their activities as well as their role in provision/protection of agri-environmental public goods it is an important pre-requisite for successful engagement of farmers into collective action.

As will be further detailed in the following section, a more pro-active engagement of farmers into collective action is also the result of an increasing involvement of farmers in the decision making process, both regarding the design and the implementation of the agri-environmental strategies to be undertaken.

6.3 The Provision of Public Goods Through Collective Action

A crucial issue in the debate on agriculture and public goods is related to the mechanisms that lead to the provision of such goods through collective action and to the cost-effectiveness of a collective approach to delivery compared to individual action.

As described in Chap. 1, the current debate on the re-orientation of the CAP towards the provision of public goods and services is largely rooted in a conventional conceptual framework, based on neo-classical economics, where the public goods provision is strongly related to the concept of jointness of production and of externalities.

According to this approach, in case of 'market failure', public intervention is needed to avoid the under provision of goods which present a high level of non-excludability and non-rivalry.

By carrying out their analysis on public goods provision through an institutional approach, some authors have shown that this conventional approach presents several conceptual and operational limitations, mainly related to the narrow definition of private/public goods (Kaul and Mendoza 2003; Hagedorn 2008).

According to these scholars, while the definition of public goods based on (non)excludability and the (non)rivalry is based solely on market criteria, the private and public domains of goods are also determined by the general public and by the political process. Thus, it is necessary to consider goods not only in their original forms, but also as social constructs and as results of deliberative policy choices (Kaul and Mendoza 2003). Indeed, together with the evolution of social institutions, many goods have developed into mixed cases, showing that exclusive and non-exclusive proprieties have a temporary dimension, since they may change, for example, as result of the development of new technologies and according to the different policies and regulations that are implemented.

As highlighted by Paavola and Adger (2005, p. 364), 'the economic approach has been silent on implementation because it conflates all policy concerns to the choice of the policy instruments. By contrast, the institutional approach sheds light on the compatibility of governance solutions and patterns of interdependence as well as on the transaction cost implications of the institutional design of governance solutions'.

From this perspective, even though the need for state intervention is inevitable in a lot of cases, the present research shows that rural stakeholders can be also encouraged to create innovative institutional arrangements and innovative territorial strategies themselves, in order to develop new possibilities to remunerate the farmers for the provision of non-commodity outputs.

The main assumption of this research is that agri-environmental issues and their solutions are 'socially built': they do not exist a priori but the participants involved in defining the problems construct them. Thus, in order to assess the role of collective action in public goods provision, multiple units of observation were considered, including farmers, public and private farmers' advisors, representatives of local and regional agencies and independent experts.

In the two case studies analyzed here, different public goods were involved and above all, different dynamics for their provision were observed, based on the different institutional and territorial contexts where collective action was implemented.

At the same time, the analysis of the two examples of collective action could improve the knowledge of the territorial dimension of the public goods involved, by giving some insights into the institutional and social mechanisms involved, as well as on the adequacy of the policy tools currently implemented.

Although the data collected in this research does not allow us to assess directly the provision of the agri-environmental public goods provided, the in depth analysis of two local initiatives allowed exploring the process beyond the provision of public goods through collective action. Here it is argued that top-down strategies (usually based on command and control measures) are based on a narrow definition of public goods (see sect. 1.4), which usually results in a strong focus on the policy tools necessary to provide/protect the public goods.

Conversely, the analysis of the provision and protection of the public goods associated through an institutional approach shows that collective action may increase the level of publicness of such goods (see Fig. 6.1).

As described in sect 1.1, the triangle of publicness provides a theoretical foundation to assess the level publicness of goods according to the following three dimensions: publicness in decision making (PD), publicness in distribution of benefits (PB) and publicness in consumption (PC). The triangle A of Fig. 6.1 is an ideal type, representing a good that is public in consumption, based on decision-making that fully meets the condition of the generalized equivalence principle, with net benefits evenly distributed across diverse population groups. As highlighted by Kaul

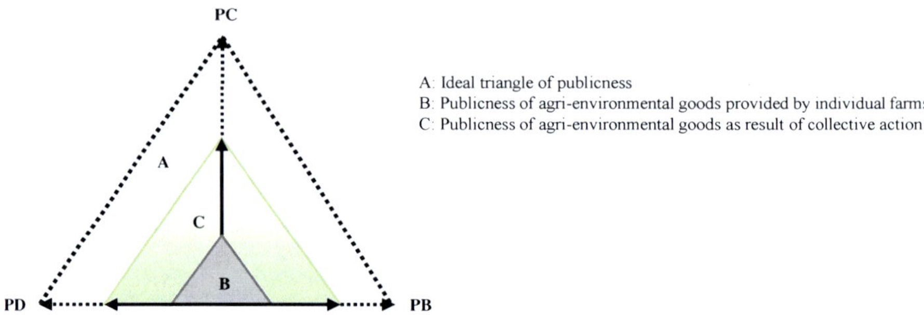

Fig. 6.1 The increased "publicness" of goods as result of collective action. (Source: author, based on Kaul and Mendoza (2003))

and Mendoza (2003), in reality, it is very unlikely that the public goods provision would meet exactly these conditions and also public goods associated to agriculture are likely to differ from this ideal triangle.

In order to assess the level of publicness of goods it would have been necessary to define and analyze clear indicators and reliable measurement methodologies for each case, and doing so was beyond the scope of this research. Nevertheless, while an increasing amount of literature shows that the level of publicness of environmental public goods which are provided by individual farms is usually low (triangle B), the present research shows how this may be increased as result of collective action.

Indeed, the case studies analyzed here, together with the main literature on the topic, show that, unlike the mainstream and top-down strategies, collective action allows increasing:

- The publicness of decision-making (PD), since collective action increases the participation of the local stakeholders in the decision-making process. Through collective action, rural stakeholders can take part in the development of territorial strategies aimed at placing goods in the public domain and in participating in the debate regarding the implementation of those strategies;
- The publicness in distribution of benefits (PB), since the participatory nature of collective action (with concerted and negotiated conditions) in many cases enables local stakeholders to derive higher benefits from the public goods provided;
- The publicness of consumption (PC), since in many cases collective action, by broadening the spatial and temporal dimension public goods provision, enables larger group of users and consumers to enjoy the benefits from these public goods.

The institutional approach adopted here considers the provision of agri-environmental public goods as a multi-actor activity, which involves social and institutional processes that go well beyond the assessment and evaluation of public goods provided. It may be argued that a stronger involvement of local actors in the decision-making process, in the distribution of benefits and in the 'consumption' of public goods leads to a reconfiguration of the modalities of implementing agri-environmental policies. More specifically, the present study shows that, compared to the conventional policy tools broadly implemented for the provision of public

Table 6.4 The public goods provided through the collective actions

Public goods	Custody of the territory	Valdaso AEA
Environmental goods	Hydro-geological management, Landscape, resilience to flooding	Increased air, soil and water quality
Socio-economical goods	Social capital, Institutional capital	Social capital, Institutional capital
Cultural and relational goods	New knowledge, capacity building, new networks	Capacity building at institutional level

goods through agriculture (incentives, control and sanctions), collective action may broaden the scope of the agri-environmental action, by facilitating a process of joint learning amongst rural stakeholders. Indeed, by focusing on specific territories rather than on single farm units, collective action involves coordination mechanisms at different levels whereby new knowledge regarding agri-environmental public goods is produced, exchanged and diffused. The dimension of joint learning is also crucial to expand the scope of the strategies implemented, by broadening the scope of the action from the agri-environmental sphere to a process that usually results in innovative institutional arrangements and in a stronger involvement of local communities. Here it has been observed how a collective approach for the provision of the environmental goods (resilience to flooding, air and water quality, etc.) also has the potential of increasing the provision of immaterial benefits as well as social and cultural goods, such as new knowledge, social capital, institutional capital and capacity building (Table 6.4).

6.4 Policy Innovation for Collective Action

What type of government intervention, both in terms of policy tools and institutional support, is necessary to promote successful collective action for the provision of agri-environmental public goods? As discussed in the first chapter of the book, the combined action of the instruments actually implemented in the framework of the CAP (cross compliance and agri-environmental measures) in many cases has been largely inefficient and, above all, shown to be inadequate in improving the provision of such goods at the required scale (Cooper et al. 2009).

It may be argued that the approach currently adopted, focused on single measures addressed to single farms, does not take into account the territorial dimensions in which farmers operate, the local environmental features and the different decision making processes regarding the agri-environmental public goods associated to agriculture.

As argued by Polman et al. (2010), in many cases agricultural innovation policies are much too focused on farms, with the main objective being the improving farmer's income, while these policies lack a clear strategy for creating and managing Common Pool Resources. The results of this research confirm that the territorial dimension of several public goods and services provided by farmers should involve an approach less based on public goods provided by single economic actors, and more on the local and regional management of natural resources.

Thus, as shown by the case studies analyzed in this book, an effective provision of public goods in rural areas should focus not only the policy tools to be implemented, but also on the social and institutional arrangements and on the collective dimensions of the decision making processes.

From this perspective, agricultural and rural development policies should give more attention to dimensions of knowledge and learning in order to activate a virtuous circle where farmers may play a pro-active role in delivering agri-environmental goods. As observed in the case studies, in many cases appropriate land management strategies depends not just on the economic incentives to land managers, but also on other factors, more related to farmers' motivations, attitudes and skills.

When analyzing the agri-environmental strategies implemented in the framework of the CAP, the difficulties of implementing collective and territorial strategies are evident (see Chap. 1). In the framework of UE agri-environmental policies, for example, the different institutional levels, the complexity of the strategies and the differences amongst territories and farming systems have prevented an extensive adoption of territorial and collective strategies for public goods provision.

Moreover, a collective approach to agri-environmental strategies has not been as widely applied because the current legal basis for rural development in many cases is not setting the necessary pre-requisite for a successful implementation of collaborative schemes. The main reasons are related to the characteristics of collective actions, which are dynamic processes, difficult to measure directly, with performance and outputs highly dependent on local institutional settings and on the social and human capital present in the different contexts.

Since collective action is targeted to local conditions, both in terms of farming systems and local institutional settings, the main barriers are related to the design of a common framework which allows the integration of collective strategies into mainstream policies.

From this perspective, as showed in the cases of collaborative AES in the Netherlands, Germany and Australia (see Chap. 3), public policies should be specifically designed and implemented to take effect on the collective dimensions, both by helping collective and territorial strategies to be developed and by supporting and enhancing territorial projects where they are already in place. These case studies, as well as other examples described in the literature (see for example Poláková et al. 2011) emphasized how these challenging objectives may be at least partially addressed through the engagement of a range of local stakeholders in the design process of the collective agri-environmental strategies. The involvement of the key local actors is important for increasing the effectiveness of the designed schemes, since it allows for the designing of measures tailored to local needs and, above all, with effective environmental outcomes.

Moreover, a fruitful interaction amongst farmers, local authorities, private sectors, advisors and local communities is also important for an effective implementation of the measures on the ground, by making the delivery of the policy measures more

Table 6.5 Policies implications: lessons from the two case studies

Policies implications	Custody of the territory	Valdaso AEA
Type of funding	Local financial resources, integrated with some support from the RDP	Project financed through coordinated set of RDP measures
Objective reached	Provision of environmental services as well as additional benefits, social and institutional capital, knowledge generation	Protection of agri-environmental public goods at territorial scale, knowledge generation
Areas	District, wide area with farmers with different characteristics: differentiated effects of the project	Delimited geographical area with well-defined farming systems (similar structure and specialization)
Coordination	Engagement of a wide group of institutions, individuals and sectors: problems of coordination	High level of coordination amongst local stakeholders that implement the agreement
RDP support	Lack of specific funding and lack of flexibility of the RDP (i.e. monitoring activities to prevent flooding could not be financed)	Increased administrative burdens and higher transaction costs
Other support needed	Further investments (time, labour and funding) for communication and advice activities	Initial capacity building process as well as funding for the coordination, management and group activities
Potentialities	This approach can bring small farmers, currently excluded from CAP support, into the system	This approach may have a strong impact on farmers knowledge and farmers' approach on innovation and training

effective (Poláková et al. 2011). This engagement in both the design and in the implementation of the schemes can help to increase the transfer and the sharing of knowledge across different individuals and groups, and this may result in the longevity of the strategies proposed.

While there are several examples of successful collaborative AES, the Italian case studies aimed at exploring grass roots and bottom up collective actions for public goods and their potentialities in scaling up and in representing a sound alternative to the conventional agri-environmental policies currently in place.

Even though these initiatives are related to specific social, economic and environmental dimensions, they provide useful insights on the potentialities and barriers to further support these grass roots collective actions through agricultural and rural development policies. The table below summarizes the main policy implications that have been observed through the analysis of the Italian case studies (Table 6.5).

In the case of the 'Custody of the Territory' project, the local agency has tried to finance a local and integrated strategy through local funding with the aim of improving the hydro-geological management of the district but also to support local farmers and to reduce the abandonment of the most marginal and isolated areas.

This strategy was mainly financed through local funding coming from the local reclamation tax and, for larger intervention works, through the regional RDP (measure 226). With regard to the EU policy support, a lack of flexibility of RDP has been observed, especially regarding the possibility of focusing the intervention on immaterial services, such as the monitoring activities, compared to the intervention based on investments. Moreover, the bureaucratic and administrative barriers in place have halted this initiative to be efficiently managed at a broader scale, since a more effective coordination mechanism and a stronger and more coordinated devolution to local agencies regarding the implementation of RDP would have been necessary.

On the other hand, the integrated delivery in Valdaso AEA, is based on a package of measures from the regional RDP, namely the agri-environmental measure (214) and, as an accompanying measure, the vocational training and advice measure (111).

In the two case studies a lack of flexibility and autonomy was also observed in managing the RDP funds at the local level, and this was identified as one of the main constraints, since a sub-regional level implementation of the RDP could have facilitated a more effective coordination at territorial scale. Another important constraint in the case of Valdaso AEA is related the higher transaction costs associated to this collective strategy, related to the financing of the initial capacity building process as well as funding for the coordination, management and group activities.

Moreover, RD policies usually lack in the flexibility to efficiently support spontaneous and endogenous initiatives, also because they are focused on administrative borders and are not tailored to the territorial provision of agri-environmental public goods, which usually involve a better tailoring to the natural boundaries of territories. From this perspective the present research highlights the role of local public bodies and institutions as key promoters and coordinators of specific projects related to agri-environmental public goods provision and the need, for implementing successful collective initiatives, of a stronger devolution of power and responsibilities. The devolution of powers to local bodies is discussed as a crucial aspect, not only in relation to the collective strategies, but also more generally for increasing the effectiveness of rural development policies on the ground. As argued by Mantino (2010, p. 15) 'better targeting and policy effectiveness are often connected with a process of devolution to local bodies and partnership and [at the same time] . . . devolution of RDP does not always represent a step forward in the improvement of policy delivery. In fact, devolution in some cases is only partial, due to the fact that marginal financial resources are devolved to sub-national bodies'.

It may be argued that this devolution, with the corresponding increasing financial resources and responsibilities, is particularly important when complex and integrated policies are implemented, such as territorial and collective action.

In the case of Custody of Territory project, for example, the Mountain Community for Garfagnana area was not in charge of the local implementation of the Regional RDP. But this is also the case for Valdaso AEA, where the provincial authority acted as intermediary between the local stakeholders and the regional government without adequate financial resources to carry out this task. A stronger involvement of those local institutions in the regional RDPs would probably have enabled local actors to elaborate more complex and articulated strategies for their territories. Thus, this programming approach to rural development should also be reinforced in order to set a strategic vision more focused on specific priorities and quantified objectives, also through a stronger involvement of regional and sub-regional actors both in the design and delivery of projects (Mantino 2011).

From this perspective, according to the proposal on rural development policies of the European Commission (2011) for the programming period 2014–2020, it is likely that the policies will be more oriented towards supporting territorial priorities, also through collective projects and strategies for agri-environmental public goods. Indeed, the RDP structure in axis will be re-shaped around thematic priorities, with an increasing focus on networks and territorial approaches, also to deliver environmental public goods. Article 36 of this proposal, for example, emphasizes how 'support for collective approaches to environmental projects and practices should help to provide greater and more consistent environmental and climate benefits than can be delivered by individual operators acting without reference to others (for example, through practices applied on larger unbroken areas of land)'.

In this new formal position for collective action, the European Commission mentions 'groups of farmers' as potential applicants and beneficiaries for the agri-environmental part of the proposal for rural development. The proposal also mentions the possibilities of supporting co-operative action, including the organization costs. Even though this may be considered a good starting point to re-orient the agri-environmental policies, the shift towards a greater emphasis on collective action of the policies should not merely be aimed at addressing the higher transaction costs that are usually associated to territorial strategies, but it should involve a deeper shift on the way policies are thought out, designed and implemented. The mainstream policies for agri-environmental public goods (such as cross-compliance, greening and agri-environmental measures) are still based on excessively 'schematized' practices with high standardization of schemes, which usually result in a tight focus on the management agreement and on the cost-effectiveness of the measures. Conversely, policies for agri-environmental public goods through collective action should increase their capacity and develop learning communities where common interests and common goals play a central role.

Finally, the present research calls for more attention to the implementation of multi-goal policy instruments through the integration of policy tools based on compensation with incentives focused on information, communication, skills and learning. Indeed, while the generation of new knowledge, social and institutional capital are among the most important positive outcomes of the collective action, such dimensions are often forgotten when planning the delivery of agri-environmental public goods in rural areas. Collective action may result in a long-term shifting in farmers' thinking and action, which ensure that public goods provision is not dependent on the financial resources available but on the local institutional and social capital.

References

Berkes F (2009) Evolution of co-management: role of knowledge generation, bridging organizations and social learning. J Environ Manage 90(5):1692–1702

Black AW (2000) Extension theory and practice: a review. Aust J Exp Agric 40:493–502

Cooper T, Hart K, Baldock D (2009) The Provision of public goods through agriculture in the European Union. Institute for European Environmental Policy, London

Deuffic P, Candau J (2006) Farming and landscape management: how French farmers are coping with the ecologization of their activities. J Agric Environ Ethics 19(6):563–585

European Commission (2011) Proposal for a Regulation of the European Parliament and of the Council establishing rules for direct payments to farmers under support schemes within the framework of the common agricultural policy. Brussels, 19.10.2011 COM (2011) 625 final/2 2011/0280 (COD)

Hagedorn K (2008) Particular requirements for institutional analysis in nature-related sectors. Europ Rev Agric Econ 35(3):357–384

Kaul I, Mendoza RU (2003) Advancing the concept of public goods. In: Kaul I, Conceição P, LeGoulven K, Mendoza RU (eds) Providing Global Public Goods. UNDP, New York

Mantino F (2010) Understanding delivery mechanisms in EU rural development policies: an institutional approach. Paper presented at the 118th Seminar of the EAAE "Rural Development: governance, policy design and delivery". Ljubljana, Slovenia August 25–27

Mantino F (2011) Developing a territorial approach for the CAP. A discussion paper, Institute for European Environmental Policy, London

Paavola J, Adger WN (2005) Institutional ecological economics. Ecol Econ 53(4):353–368

Petrick M, Gramzow A (2012) Harnessing communities, markets and the state for public goods provision: evidence from post-socialist rural Poland. World Devel 40(11):2342–2354

Poláková J, Tucker G, Hart K, Dwyer J, Rayment M (2011) Addressing biodiversity and habitat preservation through measures applied under the common agricultural policy. Report prepared for DG agriculture and rural development, Contract No. 30-CE-0388497/00-44. London: Institute for European Environmental Policy

Polman N, Poppe KJ, Schans JW van Der, Ploeg JD van der (2010) Nested markets with common pool resources in multifunctional agriculture. Rivista di Economia Agraria LXV(2):295–318

Pretty J (2003) Social capital and the collective management of resources. Science 302:1912–1914

Vanclay F, Lawrence G (1995) The environmental imperative: eco-social concerns for Australian agriculture. Central Queensland University Press, Rockhampton

Chapter 7
Conclusions

Abstract The theoretical foundation of the European agri-environmental policies, based on neo-classic economics, in many cases fails to take into adequate consideration the territorial and collective dimensions of public goods associated to agriculture. The institutional economics approach, conversely, may provide a more comprehensive and integrated basis on which to develop adequate forms of intervention, based on innovative institutional arrangements that may determine a pro-active engagement of rural stakeholders in the provision of environmental goods and services.

The book explored the multifunctional role agriculture though this institutional perspective and analyzed the role of collective action as a viable solution to provide and protect agri-environmental public goods at territorial scale. The research shows that the success of collective action is highly dependent on the capacities of local actors in stimulating social, cultural and institutional innovation. The CAP should take into more account innovative the agri-environmental strategies based on collective action which, by stimulating participation and social learning, have the potential of delivering a broad range of public goods and services that the government is unable to deliver itself.

Keywords Institutional economics · Public goods · Collective action · Common agricultural policy

Agriculture provides food as well as environmental public goods such as agricultural landscape, biodiversity, flood and drought control and carbon sequestration. The concept of multifunctionality has represented the main analytical framework, both in the institutional and academic debate, to describe the complex relations between agriculture and these public goods.

At the same time agriculture not only provides public goods, but it also produces negative externalities, such as air and water pollution, soil erosion, for example as a result of an excessive use of fertilizers of chemical inputs and of other unsustainable farming methods.

In order to incentivize the multifunctional role of agriculture and to overcome the market failures caused by public goods and externalities, the main measures adopted in the framework of the CAP are cross compliance and agri-environmental measures.

In this book it is argued that the theoretical foundation of these policy tools, based on neo-classic economics, in many cases fails to take into consideration in an adequate way the social construction of the public goods that may be provided

F. Vanni, *Agriculture and Public Goods*,
DOI 10.1007/978-94-007-7457-5_7, © Springer Science+Business Media Dordrecht 2014

by farmers and by other stakeholders in rural areas. A narrow focus on policy tools that aim at modifying the land management practices at farm level leaves behind the territorial and collective dimensions of the strategies implemented, by resulting in an unsatisfactory recognition of the knowledge and competencies needed of the rural stakeholders in order to effectively implement the policy tools proposed.

These conceptual limitations surrounding the conventional approach were described through the analysis of the most relevant studies which have addressed the provision of public goods through alternative approaches, mainly rooted in institutional economics (Kaul and Mendoza 2003; Hagedorn 2008; Ostrom 1990). These authors have observed how an institutional approach in studying public goods issues may provide a more comprehensive and integrated basis on which to develop adequate forms of intervention, based on specific policy tools but also on innovative institutional arrangements and governance patterns that can be effective at the local level.

By applying this approach to the provision of agri-environmental public goods associated to agriculture, the research shows the need for analyzing the multifunctionality of agriculture from a more integrated and wider perspective, by taking into better account the territorial, social and economic contexts where farmers operate, since the extent to which agri-environmental public goods may be provided is highly influenced by a wide range of—often interrelated—institutions and policies. For this reason it is necessary to go beyond the analysis of livelihood strategies, by exploring the institutional arrangements necessary to implement successful strategies which enhance the multifunctional role of agriculture at a broader scale. According to this approach, 'jointness of production' is not defined in relation to the production functions, but with regard to the economic and institutional arrangements which may deliver a combination of commodities and non-commodities, private and public goods (Hagedorn et al. 2002).

The book explored this new perspective on multifunctional agriculture by analyzing the role of collective action in the provision of agri-environmental public goods.

It is increasingly recognized that public goods such as biodiversity and landscape may be provided efficiently only by multiple persons and through the collective action of farmers and other stakeholders in rural areas. This approach emphasizes the need for widening the action beyond the individual farm, by adopting a more integrated approach towards achieving sustainable solutions in rural areas.

One of the main goals of the book was to explore the internal dynamics of collective action, in order to understand how the problems of public goods provision through agriculture were formulated at the local level, by observing how stakeholders express their views and motivations, and by identifying what resources were mobilized during the process. The participatory methods used in the research seemed particularly appropriate to pursuing these objectives, since through participatory tools it was possible to gain a deep understanding of the local collective action under study by exploring the relationships amongst the different units and identifying the main feedback mechanisms.

Thus, Rapid Rural Appraisal methods were used to carry out two case studies in Central Italy, where two collective actions were promoted to increase the provision of agri-environmental goods and services at territorial level. Through semi-structured interviews with local stakeholders and transect walks with farmers it was possible to explore the main mechanisms related to the decision process as well as the attitudes, motivations and social norms beyond these initiatives.

The case study "Custody of the Territory", by analyzing an innovative scheme of Payment for Environmental Services (PES) developed by a local government agency in Tuscany, has highlighted the important role of co-production (of knowledge and services) in collective action. Innovative institutional arrangements amongst rural stakeholders focused on co-production may have the potential of enhancing the multifunctional capacities of agriculture and of delivering environmental services in a cost-effective way. Moreover, the results of the analysis show that this holistic approach of delivery adopted by the local agency has the potential to address in a more coherent and integrated way the social and environmental needs of mountain areas.

The case study related to the AEA in the Valdaso area, on the other hand, demonstrated that a territorial approach for public goods provision through integrated delivery may represent an efficient and innovative way to protect environmental goods at a territorial scale. This is particularly interesting also because the agreement was implemented in an area characterized by intensive farming production, while a territorial approach to agri-environmental measures is mainly exploited in territories like Natura 2000 areas or regional and national parks, due to the difficulties to push farmers located in intensively cultivated areas to voluntarily join environmental strategies and agreements (Mantino 2011).

In spite of the main barriers related to the development of grass roots collective action (free riding, transaction costs, lack of specific financial and institutional support), these examples show that it is possible creating a different discourse between multifunctional agriculture and the new need of the society, through the achievement of a (institutional and social) transition processes, when new experiences are conceived, designed and legitimated.

To summarize, the research has shown that, in order to favor and support the development of successful collective action for public goods provision, it would be necessary to take further into account the following dimensions: (i) the human dimension, by focusing more on the motivation and attitudes of farmers and of the other stakeholders involved, (ii) the learning dimension, by incentivizing a collective approach of public goods provision through knowledge sharing and learning opportunities and (iii) the participatory dimension, by involving more directly the farmers into the local decision making process and by adopting innovative institutional arrangements at the local level, such as co-management and co-production.

With regard to the human development aspects, as argued by Paavola and Adger (2005:362), the theory of collective action makes several contributions to the implications of behavioral goals for collective action and choice, by showing the institutional framework and the particular decision rules related to these choices. From this perspective, this research shows that motivations beyond the participation of farmers to collective action are not only linked to the economic sphere, but also to other dimensions, which are important 'joint products' of collective action (Ayer 1997). Indeed, in many cases, the sense of belonging to their own community leads farmers to comply with norms that diverged from purely profit-oriented approaches. Farmers' behaviors in many cases are determined by their ideas, values and biases. The case studies show that public institutions, also through incentives, may also contribute to increasing this sense of belonging of farmers to their communities and may reinvigorate their willingness to participate in collective approaches. Moreover, when collective actions are effectively supported by institutions, it can contribute to reframing farmers' behaviors, by stimulating them to adopt new techniques and change the perception of their own activities and practices.

Moreover, the case studies also show that the success of collective initiatives is highly dependent on the capacities of the local actors (local institutions, farm advisory systems) in influencing collective behaviors, by structuring a range of incentives, capacity building programs and technical assistance which allows for the aligning of individual and group objectives and, above all, to overcome the contradictions between the private/public goods provision in farming activities.

With regard to the learning dimensions, the interviews and the transect walks allowed me to explore in depth the learning processes involved and the exchanges of information amongst farmers and between farmers and extension services: how decisions were arrived at with regard to particular practices; how farmers responded to advice; and whether there had been agreement, conflict, misunderstanding or negotiation about the advices which farmers receive or exchange.

The research highlights how the provision of public goods through agriculture implies new types of knowledge and learning environments, where the role of innovation is crucial. Such innovation is usually associated to technological innovation but the case studies show that it is also necessary to consider other approaches to innovation such as new regulations, new behaviors, cultural and institutional change.

Thus, when analyzing the provision of public goods through agriculture it is necessary to explore how innovation may cover many aspects of farming, both at structural level (farm structure, human resources, type of production, farm specialization), at organizational level (logistics, administration, communication issues) but also at the personal level, regarding the attitudes, motivation and learning of new practices by farmers. From this perspective, as observed by Knickel et al. (2009), more research is needed on institutional arrangements, and factors that support or hinder the diffusion and adoption of innovations: 'the role of organizations facilitating innovation as well as public innovation policies are critically important research questions that have not yet been tackled adequately' (ibid. p. 143).

The book also shows that the integration of different types of knowledge (farmers, experts and academic) is not only better for dissemination, but may also bring better solutions in the field. In terms of public goods and environmental services provided by farmers, social learning and co-production of knowledge are necessary processes to overcome the contradictions between sectionalized forms of knowledge, especially between the private/public goods provision through farming activities. Indeed, co-production of knowledge and social learning may determine a pro-active engagement of farmers in the provision of environmental goods and services.

Thus, territorial policies oriented to the provision of public goods should focus more on the implementation of adequate advisory and knowledge systems for farmers, which should be reformed with a shift from private goals (such a productivity increase) to public goals (such as environmental care).

From a policy development perspective, the broader definition of public goods proposed in this study also aimed at exploring the innovative institutional arrangements and tools which may be affective on supporting collective strategies for public goods provision. This underpins new governance dynamics, more consistent in placing the focus on the emerging agri-environmental problems as are perceived by the local stakeholders. This is a complex and challenging task, since it requires a careful weighing up of the interests of all the actors involved, but it is a viable solution to solving complex problems related to the management of environmental resources associated to agriculture. A new governance philosophy, based on co-management and co-production, may enable local governments and farmers to elaborate effective and long-term strategies. This also implies a radical change on the design and implementation of the action for public goods, since the objectives to be achieved should take the local situation as a starting point, by incorporating the wishes, ideas and capabilities of local communities, by using persuasive power instead of obligations and, above all, focusing on learning processes within the design and the implementation of policies.

From this perspective, it was observed that the approach of agri-environmental measures proposed by the European Commission in the framework of the CAP in many cases does not take into consideration in an adequate way the collective dimension of the environmental problems concerned and the decision making processes related to the strategies to be implemented. In many cases a narrow focus on policy tools aiming at modifying the land management practices at farm level leaves behind the territorial dynamics, which are highly dependent on the social and human capital. This leads to an unsatisfactory recognition of the role that innovative institutional arrangements may have in providing agri-environmental public goods at a territorial scale.

The broader definition of public goods proposed here aimed at exploring the complexities of the problem to be approached by the government at different levels. The innovative strategies analyzed, even though they cannot always be replicated in different contexts, demonstrated that it is possible to experience new types of cooperation between government and farmers and, in some ways, between government and society. Indeed, when government is capable of focusing on society's learning capacity it may reach important public objectives. This requires effective governance coordination at different levels but also a more dynamic approach, whereby fast and continued feedback on the results achieved and the barriers encountered may guarantee the transparency of the whole process.

The collective actions described in this book also show that it is possible to improve the capacity of the rural stakeholders in proving and protecting high valued public goods by creating greater synergy between the action of local citizens and the action of governments, where the development of new social and institutional processes are efficiently stimulated.

From this perspective, even though regulation will remain a crucial instrument for public goods associated to agriculture, it may be argued that the policy rules currently implemented in the framework of the CAP in many cases do not stimulate learning to increase pro-active engagement of farmers in providing public goods. On the other hand, agriculture and rural development policies should increase their effectiveness to stimulate learning by taking more advantage of the new information systems and web tools available, as well as by making full use of the unprecedented learning ability of our society, including farmers and citizens living in rural areas. An increasing access to information may stimulate new forms of collaboration between the public and the private sector, by stimulating collective action aimed at delivering public goods and services that the government is unable to deliver itself.

Indeed, while the conventional approach on public goods provision is based on hierarchical administrations, a collective approach aims for an open and horizontal form of governance to increase the provision of public goods in rural areas. Governments do not rely on absolute control, but should rather place more emphasis on releasing energy, on learning ability, and on the use of dynamic systems of regulation, at all levels, from local to global (Hajer 2011: 43).

From a policy development perspective, the future CAP, in order to provide public goods more efficiently at territorial level, should further take into account innovative agri-environmental strategies focused on specific territories, but should also figure out the modalities of up-scaling and replicating this approach in order to support to a larger extent these territorial and collective initiatives. The CAP should have a stronger focus on institutional arrangements and governance structures at local level, especially in order to support more efficiently formal and informal organizations that coordinate collective action for shared interests and public goods. However, in order to achieve an effective re-orientation of agricultural and rural development policies, it would be necessary to explore in more detail the conditions that may promote successful collective action, the related costs and benefits, and their effectiveness in addressing the different agro-environmental issues.

More exhaustive and comprehensive research on the role of collective action for public goods provision in agriculture should be carried out in order to address the same research questions in different territorial contexts across Europe. The successful initiatives studied here suggest the need to further explore collective action in order to design and test alternative agri-environmental strategies that allow farmers to engage, to innovate and to use their local knowledge to provide high valued agri-environmental public goods.

References

Ayer HW (1997) Grass roots collective action: agricultural opportunities. J Agr Resour Econ 22(1):1–11

Hagedorn K, Artz K, Peters U (2002) Institutional arrangements for environmental co-operatives: a conceptual framework. In Hagedorn K (ed) Environmental cooperation and institutional change: theories and policies for European agriculture. Elgar, Cheltenam

Hagedorn K. (2008) Particular requirements for institutional analysis in nature-related sectors. Eur Rev Agric Econ 35(3):357–384

Hajer M (2011) The energetic society. In search of a governance philosophy for a clean economy. PBL Netherlands Environmental Assessment Agency, The Hague

Kaul I, Mendoza RU (2003) Advancing the concept of public goods. In Kaul I, Conceição P, LeGoulven K, Mendoza RU (Eds) Providing Global Public Goods. UNDP, New York

Knickel K, Brunori G, Rand S, Proost J (2009) Towards a better conceptual framework for innovation processes in agriculture and rural development: from linear models to systemic approaches. J Agr Educ Ext 15(2):131–146

Mantino F. (2011) Developing a Territorial Approach for the CAP. A discussion paper, institute for European Environmental Policy, London

Ostrom E (1990) Governing the Commons; The Evolution of Institutions for Collective Action. Cambridge University Press, New York

Paavola J, Adger WN (2005) Institutional ecological economics. Ecol Econ 53(4):353–368

Chapter 8
Appendix: Methodology

Abstract The appendix provides a detailed description of the methodology employed in the research. It introduces the epistemological approach, which is based on social constructivism, and the participatory methods used by describing their relevance to address the research objectives. It also provides a detailed description of the Rapid Rural Appraisal methods utilized for the analysis of the collective action through the two case studies. Since one of the most important methodological tools used in the research has been semi-structured interviews, the final section of this chapter describes the analytical approach that has been used to analyze the data collected through the interviews, which is based on the Grounded Theory principles.

Keywords Participatory methods · Rapid rural appraisal · Semi-structured interviews · Grounded theory

8.1 Epistemological Approach

The epistemological approach used in this research is in accordance with the explorative character of the book, which aims at exploring collective action as an alternative and more effective strategy for addressing agro-environmental problems. This objective has called for the need for re-conciliating pragmatism with the necessity to broaden the knowledge about the social and institutional change that occurred in the collective actions under study.

The epistemological starting point used for the analysis of the actors involved in the collective action as well as the social and institutional process is *social constructivism*. This approach emphasizes the importance of culture and context for understanding what occurs in society and for constructing knowledge (McMahon 1997). This perspective is closely associated with many contemporary theories and it is based on the sociological and social psychological concept of social constructionism. Social constructionism is a school of thought introduced into sociology by Peter L. Berger and Thomas Luckmann with their 1966 book *The Social Construction of Reality*. The interest of social constructionism is to discover how individuals and groups create their perceived reality (McMahon 1997).

Social constructivism is based on specific assumptions about reality, knowledge and learning: social constructivists believe that reality is constructed through human

F. Vanni, *Agriculture and Public Goods*,
DOI 10.1007/978-94-007-7457-5_8,

activity and, for this reason, it cannot be discovered; in other words, it does not exist prior to its social invention (Kim 2001). According to this approach, individuals create meanings through their interactions with each other and with the environment they live in. Learning is then viewed as a social process and 'meaningful learning occurs when individuals are engaged in social activities' (*Ibid.*). This epistemological starting point reflects the nature of the book's aims and objectives, as it facilitates a better understanding of the variables affecting collective action (see Chap. 2).

This approach seemed suitable for addressing the relationships among different actors involved in the collective actions, as well as their knowledge on the environmental resources under study. Moreover, this epistemological approach may also be considered appropriate in analyzing the institutional arrangements related to the management of natural resources and to the development of environmental policies, which 'depend critically on the social constructions of environmental problems' (Hajer 1995:264). It must also be highlighted that the main assumption of the research is that agri-environmental issues and their solutions are 'socially built': they do not exist a priori but the participants involved in defining the problems construct them.

Following this epistemological approach, in the framework of this research, the agri-environmental public goods and services were not established *a priori*, but they were selected and identified in the context of two territorial case studies. The main focus of the research is on how local groups address negative externalities such as the over-use or inappropriate use of pesticides but also as local groups may organize (collective) action for providing or protecting public goods such as biodiversity and natural habitat preservation. The identification by local communities of environmental goods and services to be provided and/or protected is therefore crucial, since the local actors are the main stakeholders who must take (collective) action to address the local agri-environmental issues.

In order to analyze collective action for environmental goods and services, it was considered particularly relevant to explore the local dimension of this action, since the local context is where people have a practical and experiential relationship with the objects and processes that have to been managed. In addition, the local dimension allows the exploring of the local interdependencies, since required changes such as land use and management practices are most likely to emerge from action, social relations and experiences that take place locally (Steyaert et al. 2007).

As specified in the following section, in order to explore in depth the local dynamics and the social construction of the agro-environmental issues under study, perceptions, attitudes and actions of local stakeholders involved in the collective action were explored through participative research methods.

8.2 Participatory Methods: An Overview

The main features of conventional research may be resumed as there is a tendency towards packaging intervention methods and programmes into one-size fits-all, off-the-shelf approaches, and usually this approach is based on a notion of universal

Table 8.1 Methods reversal (Source: Adebo (2000))

Conventional research	PRA
Method: closed, individual, verbal, counting	Method: open, group, visual, comparing
Relationship: extracting, distance	Relationship: empowerment, rapport
Experience: boredom, frustration	Experience: fun/enjoyment, relaxed

best practices. Conversely, participatory methods address the drawbacks inherent in the conventional approach by actively involving end-users in the research process, incorporating their views and representation into the prioritization, review, conduct, and dissemination of scientific research (Lilja and Bellon 2008:479).

During recent decades an increasing acceptance and widespread use of participative research in many fields was observed. According to Campbell (2001) this is due to the following five principles: (i) optimizing trade-offs between the cost, quantity, relevance, truth, and benefit of possessing information; (ii) offsetting biases from development 'tourism' and hurried visits; (iii) triangulating by using more than one method or source to obtain and cross-check information; (iv) learning directly from with rural people whose knowledge and capacities are generally underrated but required for sustainable development; and (v) learning rapidly and progressively through flexible, iterative, interactive processes aimed at the conscious exploration of local peoples' lives.

In the context of the participative research methods, Participatory Rural Appraisal (PRA) was settled on as an approach to development planning and as a method of investigation evolved from many different sources. Some of the sources were modified in order to be utilized in a participatory mode, and the others were taken up as they were used for investigation and planning. PRA was described by Chambers (1994a:1253) as a 'growing family of approaches and methods to enable local (rural or urban) people to express, enhance, share and analyze their knowledge of life and conditions, to plan and to act'.

PRA methods aim at gaining accessibility to rural people by involving them in the research process, that is in learning, collecting, finding and analyzing information in open discussions and total interaction, with the additional result that this participatory research also increases accessibility to sites. In several ways, PRA methods reverse the conventional ways of doing research, since it assumes that local communities have knowledge and information, but it needs to be organized with resources which need to be mobilized, by integrating traditional knowledge systems and external knowledge in the research process (Table 8.1).

According to Chambers (1994a:954) there are five streams which stand out as sources and parallels to PRA: (i) activist participatory research; (ii) agro-ecosystem analysis; (iii) applied anthropology; (iv) field research on farming systems; (v) Rapid Rural Appraisal (RRA).

The RRA, in particular, is quite a widespread technique mainly used in development studies as a preliminary stage when embarking on surveys of farmers. The technique essentially involves an informal, rapid, exploratory study of a specified geographical area designed to establish an 'understanding' of local agricultural conditions, problems and characteristics. RRA may provide basic information on the feasibility of beginning a survey project in an area, particularly when one is intending to survey an area about which little is known (FAO 1997).

The philosophy, approaches and methods of RRA may differ from those of PRA. Although many of the set of practices overlap, there are several distinctions (Chambers 1994a:958–959): RRA methods are more verbal, with outsiders more active, while PRA methods are more visual, with local people more active, but the methods are now largely shared; the RRA (extractive-elicitive) approach where the main objective is data collection by outsiders, and a PRA (sharing-empowering) approach where the main objectives are variously investigation, analysis, learning, planning, action, monitoring and evaluation by insiders.

PRA aims at assisting local people to plan, implement, monitor and evaluate their own action plans. Since it aims at people taking action themselves it is most suited for the community level, PRA presents a major step forward from RRA: local people do the analysis and plans for the future (Khodamoradi and Abedi 2011:75). In any case, as observed by Chambers (1994a), a single formula of approaches and technique should be avoided, both because it may hinder the creativity during the fieldwork and because the guiding principle of PRA/RRA is to use 'your best judgment at all times'.

PRA/RRA methods do not refer to a single technique, but to a range of investigation procedures which rely mainly on experts' observation, coupled with semi-structured interviewing with farmers, local leaders and officials. These techniques have much in common with the social anthropologists' case study approach but are executed over a period of weeks, or at most months, rather than extending over several years (FAO 1997).

Chambers (1994a) provides a detailed list of techniques which may be used in the framework of PRA/RRA research. Amongst them, the most relevant techniques are:

- *Secondary sources*: analysis of files, reports, maps, photographs, academic and non-academic literature;
- *Semi-structured interviews:* open-ended interviews on a topic;
- *Group discussion and brainstorming*: focus groups by local people alone, by local people and outsiders together, by outsiders alone;
- *Do-it-yourself and Asking to be thought*: performing a particular task along with local people after it being demonstrated;
- *Transect walks*: walking with local people, observing and asking questions;
- *Seasonal calendars, daily time use analysis, and livelihoods analysis*: analysis of seasonal changes (on crops, labour, diet, migration, income, expenditures, etc.), daily changes (schedules of the main activities with also seasonal changes), general analysis of livelihoods (stability, crisis, income, multiple activity, financial constraints etc.);

- *Institutional, Chapaty or Venn diagramming*: identification of individuals and institutions important in and for a community, or within an organization and their relationships;
- *Matrix scoring and ranking*: ranking system or scoring to compare different physical characteristics (soils, tree, etc.) or different methods (soil, water conservation strategies, etc.).

All the methods listed above can be used in both RRA and PRA, but some are more emphasized in one than the other. RRA has tended to stress the use of secondary sources and verbal interaction especially through semi-structured interviewing, and observation: so these are sometimes described as 'RRA methods'. PRA methods are more focused on shared visual representations and analysis by local people, such as mapping or modelling on the ground or paper; estimating, scoring and ranking with seeds, stones, sticks or shapes; Venn diagramming; free listing and card sorting; linkage diagramming; and presentations for checking and validation (Chambers 1994a:959).

Even though PRA/RRA are mainly used from a development study perspective, these methods are applicable in a very wide range of situations, including natural resource management, agriculture, poverty, social and health care programmes, both in rural and urban areas and both in developing and industrial countries (Chambers 1994a, 1994a; Khodamoradi and Abedi 2011).

8.3 Data Collection and Analysis: Rapid Rural Appraisal (RRA)

In the context of this book, participatory research methods were applied in two case studies of collective action (Chap. 4 and Chap. 5). An approach based on case study was considered the most appropriate for the following reasons:

- It was necessary to gain a holistic perspective of the different local projects involving collective action in order to have a deep understanding of the different units, of the links and networks in place, together with the main feedback mechanisms; Territorial case studies were considered the most appropriate method to attaining the research objectives, which are related to social and individual learning, local institutions, territorial governance and local farming systems.
- A case study approach was considered appropriate since agri-environmental public goods are highly local and case-specific but also because different disciplines overlapped in the analysis and it was necessary to have a holistic approach to the phenomena under study.

By taking into consideration the general frameworks of PRA/RRA methods, in the context of the present book one of the most significant streams is agro-ecosystem analysis. This approach, by drawing on systems and ecological thinking, combines analysis of systems and system properties (productivity, stability, sustainability, and equitability) with pattern analysis of space (maps and transects), time (seasonal calendars and long-term trends), flows and relationships (flow, causal, Venn and other diagrams), relative values (bar diagrams of relative sources of income etc.), and decisions (decision trees and other decision diagrams) (Chambers 1994a:954).

Amongst all the RRA methods, in the context of the present research four methods were mainly used: (i) secondary sources; (ii) sketch maps; (iii) transect walks; (iv) semi-structured interviews and (v) institutional and system diagrams.

- *Secondary sources: literature review and document analysis.* One of the most important approaches in integrating other qualitative methodologies is documentary analysis. The documentary analysis was drawn from different bodies of documentary material: academic articles, policy documents, conferences' minutes, technical documents. Public documents (i.e., policy documents, newspaper articles, and conferences' minutes) were particularly useful in exploring the public profile of the projects under study and analysing the institutional and economic relationships between the different actors (farmers, politicians, technicians and other stakeholders). Technical studies (i.e. academic articles, policy evaluation documents, environmental analysis, monitoring reports) and other available sources have been useful for assessing the social and environmental outcomes of the collective actions under study. The use of this material has also helped to evaluate the validity of the data gathered through the interviews.Indeed, the need for triangulation arises from the ethical need to confirm the validity of the processes and increases the reliability of the data and the process of gathering it (Tellis 1997). In order to reach this aim, the documentary analysis has been crucial, by evaluating whether or not different sources of data agree with each other.
- *Sketch Maps.* The purpose of sketch maps is to provide a visual representation of what the local actor perceives as their territory. This includes showing the shape (appearance) of the community, boundary and all the major features, including where resources, activities, problems and opportunities are located, as well as the dimension and scope of issues to be investigated. This method was conceived as a convenient way to systematically and accurately record basic data of the area under study, where local actors and farmers identified the main characteristics of the territorial systems. Such features, such as major and minor communication roads, major and minor settlements (villages, towns), land use, farms, rivers, water ways, and channels were annotated to illustrate the location and extent of each of the characteristics observed.
- *Transect Walks.* A transect is a walk or a series of walks through an area with local informants to learn of the range of different conditions, problems and opportunities of the area. Transect walks, by showing a cross section of the area as observed by the walk, provide very useful mapping information beyond that collected during the initial meetings and verifies the information on the sketch map. It may also add information to specific features of the areas under study (slope drainage, vegetation, water, soils other sources). Thus, through transect walks it is possible to have more detailed and specific information on cropping patterns, trees, vegetation, farm size, problems and opportunities. Indeed, an important consideration underlying the choice for this method was seeing farmers acting in the field, see his/her enthusiasm for the nature conservation and environmental stewardships

and assessing his/her knowledge about the subject. This method was combined with the semi-structured interviews and documented in interview transcripts. In the context of the research, systematic walks were carried out with farmers and the local actors in the areas under study with the main objectives of observing, discussing, identifying different forms, local techniques, local knowledge, seeking local uses, problems, solutions and opportunities. Transect walks also allowed the full exploration of the areas concerned, the spatial and physical differences, and the assessment of the main infrastructures and activities carried out by the local actors in their everyday life.

- *Semi-structured interviews.* The data collection process was largely based on semi-structured interviews, which may be considered a suitable method for researching the motivations, attitudes and values about a given theme. The interviews were conducted on the basis of a questionnaire with open-ended questions, keeping the possibility open to discuss related topics in the course of the interview. The answers were analyzed from taped transcripts also by extracting statements that were considered particularly relevant in relation to the different issues. Due to the nature of the research and its objectives, selecting interviewees using sophisticated sampling techniques was not considered essential. The sample of farmers and local actors interviewed for each case study was selected with the objective of representing a wide cross-section of interests in the local communities. With regard to farmers, the main criteria were selecting farmers form their location on the area under study, covering both small and large farmers, farmers with a large experience on the project under study and new entry farmers. In addition to the farmers, other key informants (informants with specialized knowledge) were selected in the local communities who had extensive and deep knowledge about collective action, the rural development patterns of the concerned areas and of local conditions and problems. The use of information provided by local farmers and key informants allows the accessing of information otherwise hard to gather. The analysis of interview data has been carried out following the Grounded Theory approach (see below).
- *Institutional and system diagrams.* This technique was used in order to illustrate through graphs and diagrams the main local groups involved in each case study (including farmers, local communities, local institutions, environmentalist organizations, advisory systems staff, commercial interests, NGOs, researchers, etc.), the interactions between these groups and the interaction between the stakeholders and the resources/public goods under study (see Asia Forest Network 2002). Information on local groups and local stakeholders was collected mainly by using secondary sources and semi-structured interviews with key informants. This information was then diagrammed in a systematic way. In the institutional and system diagrams arrows and lines were used in order to indicate the various interactions between different user groups. Semi-structured interviews were particularly useful for identifying the local groups and the local stakeholders involved in the case studies and to establish the types of interaction amongst them.

8.4 Overview of the RRA methods

As showed by Meinzen-Dick et al. (2004), in order to assess the effectiveness of collective action it is necessary to analyze multiple units of observation, including individuals, groups, farms and landscape, since perceptions of different dimensions of effectiveness may well differ amongst respondents, especially amongst group leaders and other members, as they are likely to have different knowledge, viewpoints and biases.

The participatory methods in the case studies were considered particularly appropriate to exploring collective action, since it involves analyzing local phenomena, exploring the relationships amongst the different units and identifying the main feedback mechanisms.

These methods were used to learn from and with the farmers and the local actors, by learning directly, on-site, and face-to-face and by gaining directly from them the physical, technical and social knowledge. Indeed, local actors' perceptions and understanding of resource situations and problems are important to learn and comprehend, because solutions must be viable and acceptable in the local context, and also because local inhabitants possess extensive knowledge about their resource setting (FAO 1997).

The objective of the case studies was exploring the variables affecting collective action and, as described in Chap. 2, RRA methods were used to analyze:

1. Resource system characteristics;
2. Group characteristics;
3. Institutional arrangement;
4. External environment.

Through the analysis of the variables 1 and 2 it was possible to explore farmers' behaviors and the innovative farming practices associated with the agri-environmental public good(s), as well the learning and social environment in which farmers operate, together with farmers' preferences, attitudes and perceptions.

On the other hand, the variables 3 and 4 are related to institutional and policy setting, and were analyzed in order to assess the different institutional systems, the territorial strategies, and the institutional capacity relevant for the provision of agri-environmental public good(s) through collective action (Table 8.2).

8.5 Grounded Theory and Analysis of the Qualitative Data

The main sources of data and information of this study were the semi-structured interviews with the local stakeholders. By following the constructionist approach described above, the analysis of the data followed the Grounded Theory principles.

Grounded Theory (GT) is structured in a way that the researcher asks a question or series of questions which are designed to lead to the development or generation of a theory regarding some aspect of social life. GT is a qualitative approach that uses

Table 8.2 The analysis of collective action through participatory methods

Variables affecting collective action	Objectives	PRA/RRA methods
1) Resource system characteristics	Identification and analysis of agri-environmental public goods	Secondary sources
	Identification of farming knowledge and practices	Semi-structured interviews
	Analysis of farmers' attitudes, motivations and behaviors	Transect walks
2) Group characteristics	Identification of the actors involved in the collective action and their role (group size and heterogeneity)	Sketch maps
	Identification and analysis of the relationships among the involved actors (social capital and communication)	Secondary sources
		Semi-structured interviews
		Transect walks
3) Institutional arrangement	Identification of the institutions involved	Secondary sources
	Analysis of the mechanisms for control, enforcement, monitoring and evaluation	Semi-structured interviews
	Identification of new drivers, innovative decision making paradigms	Institutional and system diagrams
4) External environment	Analysis of the typology of the implemented measures and strategies	Secondary sources
	Description of the way policies are designed, implemented at the administrative level and delivered to farmers	Semi-structured interviews
	Assessment of the outcome of the policy measures and local strategies (identifications of the main constraints to achieve the full potential)	Institutional and system diagrams

a systemic set of methodological tools in order to develop a theory from the study and the analysis of the reality. The researcher, instead of starting from a theoretical assumption to be verified in the field, starts from the fieldwork, in order to observe the most relevant elements that emerge from the study of the reality, with the main objective of developing a theory on the topic he/she is studying.

The development of a theory is the result of a scrupulous process that involves the researcher moving in and out of the data collection and analysis processes, a movement that is usually called 'iteration'.

This 'iteration' allows a theory to develop with regard to the research questions and, above all, it allows the analyst to engage in a theoretical sampling process. As described by Glaser and Strauss (1967:45) 'theoretical sampling is the process of data collection for generating theory whereby the researcher jointly collects data, and analyses his data and decides what data to collect next and where to find them, in order to develop his theory as it emerges'.

In the framework of GT, the collection, coding and analysis of data are three processes that are always interrelated: while the researcher is active in collecting data at the same time he/she is also active in coding and interpreting them. This continuous process of asking questions and comparing data is functional to understanding the main unexplored issues as well as the main weakness of the analysis, and it allows the researcher to obtain a deep knowledge of the phenomena under study.

This comparative process involving joint data collection and analysis continues until the researcher reaches saturation, the point at which there are no new ideas and insights emerging from the new data.

With regard to the analytic phase, the most relevant action is the data *coding*, which is the process of defining what the data is all about. As observed by Charmaz (1995): 'unlike quantitative coding, that means applying preconceived codes to the data, qualitative grounded theory coding means *creating* the codes as you study your data. Since codes emerge from the research, this process may take the researcher to unforeseen and unexpected areas and questions'.

More specifically, this process involves three levels or types of coding:

- *Open coding*—the data are segmented or divided into similar groupings and forms. Preliminary categories of information about the phenomenon are developed;
- *Axial coding*—after the open coding, the identified categories are brought into groupings. These groupings resemble themes and are generally new ways of seeing and understanding the phenomenon under study;
- *Selective coding*—the categories and themes are organized and integrated in a way that articulates a coherent understanding or theory of the phenomenon of study.

The process of data collection and coding must be accompanied by the writing of informal analytical notes, commonly called *memos*. Memo-writing constitutes a crucial method in GT for the following reasons: (i) it prompts researchers to analyze and code data and early in the research process and (ii) it keeps the researcher involved in the analysis, by helping to increase the level of abstraction of the ideas (Charmaz 1996).

In this study the memos that resulted were particularly useful for making comparisons within the data, between data and codes, codes and categories, and new ideas were articulated from these comparisons. Moreover, in the context of the present research, as result of the coding and memo-writing processes, data was sorted and organized in a hierarchical structure, where the different categories and sub-categories were organized into index trees. This *sorting* process was carried out separately for each case study and two index trees were developed.

The final stage of the analysis of qualitative data collected through the interviews was the writing up of the results, which followed the structures of categories and sub-categories resulted by the sorting process. The table below summarizes the different stages of the GT and their utilization in this study (Table 8.3).

Table 8.3 The use of the Grounded Theory in the present book

Stages of grounded theory	
Data collection	Semi-structured interviews, transcription
Coding (open-axial-selective)	Data has been coded and then organized into a set of categories and subcategories. The process of data collection and coding continued until the point where theoretical saturation was reached
Memos	Writing of memos about the various categories and properties that emerged from the data as a result of the coding process. This process allowed the identification of links between the various themes/categories
Sorting	All the codified data and memos were organized in categories and subcategories forming two index trees (one for each case study)
Writing	The structure of the case study writing has been guided by the sorting of data above

8.6 Strengths and Limitations of the Proposed Methodology

In spite of the appropriateness of the method of case study in this research, this approach can also represent several limitations. Indeed, the validity of this study, as with any research, is determined by the extent to which its findings can be generalized. Case studies may be used as the basis for generalization, but this generalization is theoretical, and not empirical in nature (Sharp 1998:788). In order to evaluate the possibility of generalizing findings, the research has followed Mitchell's (1983) approach. Mitchell illustrates how explanations can be generated without the need for empirical typicality and emphasizes the role of atypical cases in the generation of theoretical explanations. For this reason, a likely limitation of this book can be viewed also as an important strength: the role of atypical cases (grass roots collective actions for public goods) is discovering the logical and necessary relationships among variables and they can generate theories that can be generalized (Sharp 1998:788).

With regard to the methodologies used, the qualitative nature of the research approach has represented another important limitation of the study. In fact, while the methodology adopted focused especially on the social process leading to collective action and to new forms of governance in rural areas, it is necessary to emphasize that demonstrating that collective actions which arise to address agro-environmental issues does not mean that such actions automatically lead to environmental improvements and to effective agri-environmental public goods provision.

In fact, the approach of the study, by focusing on the social construction of reality, has been limited in evaluating the material 'reality' of progress towards public goods provision (the outcomes of collective action). Since public goods provision is a function of quality as well as quantity (Meinzen-Dick 2004), the qualitative data collected in the field was also integrated with quantitative data collected through the analysis of policy and technical documents. Indeed, when trying to assess the rank of quality of different services through interviews it is clear that respondents have different 'reference' points for assessing quality, and for this reason it is necessary to integrate the data collected with quantitative data regarding the outcomes in terms of natural resource management (agricultural and environmental practices) as well as in the provision of public goods (pollution, landscape, etc.).

At the same time, the research focuses mainly both on the individual incentives and motivations of local groups to participate in collective action and on how the motivation for action can be shaped at group level (through social networks, organizations, etc.) by showing little evidence on the related bargaining powers and roles of the main actors involved in the process.

Nevertheless, in order to achieve the research objectives, the methodology tools proposed were considered flexible and appropriate to fit the local situations and to explore such issues. Indeed, the qualitative methods described above were considered particularly suitable for establishing direct rapports with the respondents, with the objective of having the insider's views and understanding of the local actors involved in the collective action.

In particular, semi-structured interviews were especially useful in terms of understanding relevant institutions as well as the rules, norms and attitudes involved in the processes under study, also allowing a cross check on information and data collected. The guideline of the questionnaire for the semi structured interviews was particularly useful in this regard, since the focus of interviews in many cases shifted rapidly and in many cases it was difficult to compare the answers from different interviews. The other Rapid Rural Appraisal (RRA) techniques such as sketch maps and transect walks were particularly useful for identifying and analyzing more in depth the practical problems of the collective actions in the field, with a deeper learning which allowed for the reduction of the gap between theory and practice.

The Interviews

Table 1A The project "Custody of the Territory" in Media Valle del Serchio (Tuscany)

Interviews	Place	Date
1. Project ideator. Ex head of Department of the Mountain Community Media Valle del Serchio	Barga (Lucca)	12th October 2011
2. Technician of the Mountain Community Media Valle del Serchio	Limestre (Pistoia)	12th September 2011
3. Town councillor ofCutigliano Municipality	Cutigliano (Pistoia)	9th August 2011
4. Project coordinator, technician of the Mountain Community Media Valle del Serchio	Borgo a Mozzano (Lucca)	14th October 2011
5. Mayor of Barga	Barga (Lucca)	12th October 2011
6. President of the Mountain Community Media Valle del Serchio	Borgo a Mozzano (Lucca)	14th October 2011

Table 1A (countinued)

Interviews	Place	Date
7. Person in charge of the agricultural sector of the Mountain Community Media Valle del Serchio	Borgo a Mozzano (Lucca)	14th October 2011
8. Software developer (IDRAMAP) HT&T Consulting	Pontedera (Pisa)	2nd September 2011
9. Software developer (IDRAMAP) HT&T Consulting	Pontedera (Pisa)	2nd September 2011
10. Farmer	Melo, Cutigliano (Pistoia)	28th March 2011
11. Farmer	Pascoso (Lucca)	1st December 2010
12. Farmer	Butale, Cutigliano (Pistoia)	9th August 2011
13. Farmer	Selvapiana Piteglio (Pistoia)	12th September 2011

Table 1B The agri-environmental agreement in Valdaso (Marche)

Interviews	Place	Date
14. Technician of ASSAM (Project Leader)	Castel di Lama (Ascoli Piceno)	15th September 2011
15. Person in charge of the Valdaso AEA for Ascoli Provincial Government	Ascoli Piceno	15th September 2011
16. Person in charge of the Valdaso AEA for Fermo Ascoli Provincial Government	Ascoli Piceno	15th September 2011
17. Ex Councillor (agricultural sector) of Ascoli Provincial Government	Roma	7th July 2011
18. Head of Agriculture Department of Marche Regional Government	Ancona	25th July 2011
19. Technician, in charge of chemical analysis, ASSAM JESI	Jesi	16th September 2011
20. Technician, ASSAM—Agrometeo Center Ascoli Piceno	Ascoli Piceno	29th November 2011
21. Technician, ASSAM—Agrometeo Center Ascoli Piceno	Ascoli Piceno	29th November 2011
22. Technician, ASSAM—Agrometeo Center Ascoli Piceno	Ascoli Piceno	29th November 2011
23. Farmer, Leader Nuova Agricoltura	Montalto delle Marche (Ascoli Piceno)	15th September 2011
24. Farmer	Montalto delle Marche (Ascoli Piceno)	15th September 2011
25. Farmer	Montalto delle Marche (Ascoli Piceno)	29th November 2011
26. Farmer	Montalto delle Marche (Ascoli Piceno)	29th November 2011
27. Farmer	Altidona (Ascoli Piceno)	28th November 2011
28. Farmer	Montedinove (Ascoli Piceno)	28th November 2011

References

Adebo S (2000) Training manual on participatory rural appraisal. http://www.myfirecommunity.net/discussionimages/NPost8220Attach1.pdf. Accessed 18 Nov 2010

Campbell JR (2001) Participatory rural appraisal as qualitative research: distinguishing methodological issues from participatory claims. Hum Organ 60(4):380–390

Chambers R (1994a) The Origin and Practice of Participatory Rural Appraisal. World Dev 22(7):953–969

Chambers R (1994b) Participatory rural appraisal (PRA): Analysis of Experience. World Dev 22(9):1253–1268

Charmaz K (1995) Grounded theory. In Smith JA, Harré R, Langenhove LV (eds) Rethinking methods in psychology. SAGE Publication, London

Charmaz K (1996) Constructing grounded theory. A practical guide through qualitative analysis. SAGE Publication, London

FAO (1997) Marketing research and information systems, marketing and agribusiness texts—4. Food and agriculture organization of the United Nations, Rome

Glaser BG, Strauss AL (1967) The discovery of grounded theory: strategies for qualitative research. Aldine Transaction, Brunswick

Hajer M (1995) The politics of environmental discourse: Ecological modernisation and the policy process. Clarendon press, Oxford

Khodamoradi S, Abedi M (2011) Review required before participatory rural appraisal (PRA). J Am Sci 7(3):74–81

Kim B (2001) Social constructivism. In Orey M (ed) Emerging perspectives on learning, teaching, and technology. Department of educational psychology and instructional technology, University of Georgia

Lilja N, Bellon M (2008) Some common questions about participatory research: a review of the literature. Dev Pract 18(4–5):479–488

McMahon M (1997) Social Constructivism and the World Wide Web—A Paradigm for learning. Paper presented at the ASCILITE conference. Perth, Australia

Meinzen-Dick R, Di Gregorio M, McCarthy N (2004) Methods for studying collective action in rural development. Agr Syst 82(3):197–214

Mitchell JC (1983) Case study and situational analysis. Sociol Rev 31(2):187–121

Sharp K (1998) The case studies in nursing research: the problem of generalization. J Adv Nurs 27(4):785–789

Steyaert P, Barzman M, Billaud JP, Brives H, Hubert B, Ollivier G, Roche B (2007) The role of knowledge and research in facilitating social learning among stakeholders in natural resources management in the French Atlantic coastal wetlands. Environ Sci Policy 10(6):537–550

Tellis W (1997) Application of a case study methodology. The qualitative report. The qualitative report, Volume 3, Number 3, September, 1997. http://www.nova.edu/ssss/QR/QR3–3/tellis2.html. Accessed 8 Nov 2011

Index

F. Vanni, *Agriculture and Public Goods*,
DOI 10.1007/978-94-007-7457-5,

MIX
Papier aus verantwortungsvollen Quellen
Paper from responsible sources
FSC® C105338

Printed by Books on Demand, Germany